Functional Design Errors in Digital Circuits

Lecture Notes in Electrical Engineering
Volume 32

For other titles published in this series, go to
www.springer.com/series/7818

Kai-hui Chang · Igor L. Markov · Valeria Bertacco

Functional Design Errors in Digital Circuits

Diagnosis, Correction and Repair

Dr. Kai-hui Chang
University of Michigan
Dept. Electrical Engineering &
Computer Science
Ann Arbor MI 48109-2122
USA
changkh@umich.edu

Dr. Valeria Bertacco
University of Michigan
Dept. Electrical Engineering &
Computer Science
Ann Arbor MI 48109-2122
USA
valeria@umich.edu

Dr. Igor L. Markov
University of Michigan
Dept. Electrical Engineering &
Computer Science
Ann Arbor MI 48109-2122
USA
imarkov@umich.edu

ISBN: 978-1-4020-9364-7 e-ISBN: 978-1-4020-9365-4

DOI 10.1007/978-1-4020-9365-4

Library of Congress Control Number: 2008937577

Printed on acid-free paper

9 8 7 6 5 4 3 2 1

springer.com

To the synergy between science and engineering

Contents

List of Figures

List of Tables

Preface

The dramatic increase in design complexity of modern circuits challenges our ability to verify their functional correctness. Therefore, circuits are often taped-out with functional errors, which may cause critical system failures and huge financial loss. While improvements in verification allow engineers to find more errors, fixing these errors remains a manual and challenging task, consuming valuable engineering resources that could have otherwise been used to improve verification and design quality. In this book we solve this problem by proposing innovative methods to automate the debugging process throughout the design flow. We first observe that existing verification tools often focus exclusively on error detection, without considering the effort required by error repair. Therefore, they tend to generate tremendously long bug traces, making the debugging process extremely challenging. Hence, our first innovation is a bug trace minimizer that can remove most redundant information from a trace, thus facilitating debugging. To automate the error-repair process itself, we develop a novel framework that uses simulation to abstract the functionality of the circuit, and then rely on bug traces to guide the refinement of the abstraction. To strengthen the framework, we also propose a compact abstraction encoding using simulated values. This innovation not only integrates verification and debugging but also scales much further than existing solutions. We apply this framework to fix bugs both in gate-level and register-transfer-level circuits. However, we note that this solution is not directly applicable to post-silicon debugging because of the highly-restrictive physical constraints at this design stage which allow only minimal perturbations of the silicon die. To address this challenge, we propose a set of comprehensive physically-aware algorithms to generate a range of viable netlist and layout transformations. We then select the most promising transformations according to the physical constraints. Finally, we integrate all these scalable error-repair techniques into a framework called *FogClear*. Our empirical evaluation shows that FogClear can repair errors in a broad range of designs, demonstrating its ability to greatly reduce debugging effort, enhance design quality, and ultimately enable the design and manufacture of more reliable electronic devices.

This book is divided into three parts. In Part I we provide necessary background to understand this book and illustrate prior art. In Part II we present our FogClear methodologies and describe theoretical advances in error repair, including a counterexample-guided error-repair framework and signature-based resynthesis techniques. In Part III we explain different components used in the FogClear flow in detail, including bug trace minimization, functional error diagnosis and correction, an incremental verification system for physical synthesis, post-silicon debugging and layout repair, as well as methodologies for spare-cell insertion. Finally, we conclude this book and summarize our key techniques in the last chapter.

PART I

BACKGROUND AND PRIOR ART

Chapter 1

INTRODUCTION

Most electronic devices that we use today are driven by Integrated Circuits (ICs) – these circuits are inside computers, cellphones, Anti-lock Braking Systems (ABS) in cars, and are sometimes even used to regulate a person's heartbeat. To guarantee that these electronic devices will work properly, it is critical to ensure the functional correctness of their internal ICs. However, experience shows that many IC designs still have functional errors. For instance, a medical device to treat cancer, called *Therac-25*, contained a fatal design error which overexposed patients to radiation, seriously injuring or killing six people between 1985 and 1987 [89]. The infamous *FDIV* bug in the Intel Pentium processors not only hurt Intel's reputation but also cost Intel 475 million dollars to replace the products [146]. A more subtle design error may alter financial information in a bank's computer or cause a serious accident by starting a car's ABS unexpectedly. To address these problems, enormous resources have been devoted to finding and fixing such design errors. The process to find the design errors is called *verification*, and the process to repair the errors is often called *debugging*. Error repair involves diagnosing the causes of the errors and correcting them.

Due to the importance of ensuring a circuit's functional correctness, extensive research on verification has been conducted, which allows engineers to find bugs more easily. However, once a bug is found, the debugging process remains mostly manual and ad hoc. The lack of automatic debugging tools and methodologies greatly limits engineers' productivity and makes thorough verification more difficult. To automate the debugging process, we propose new methodologies, tools and algorithms in this book. In this chapter, we first describe the current circuit design trends and challenges. Next, we briefly review existing solutions that address the challenges and point out the deficiency in current solutions. We then provide an outline of our approach and summarize the key techniques of this work.

1.1 Design Trends and Challenges

Modern circuit designs strive to provide more functionalities with each product generation. To achieve this goal, circuits become larger and more complicated with each generation, and designing them correctly becomes more and more difficult. One example that shows this trend is Intel's microprocessors. The 80386 processor released in 1985 barely allows the execution of the Windows operating system and contains only 28 thousand transistors. On the other hand, the Core 2 Duo processor released in 2006 supports very complicated computations and is several hundred times more powerful than the 80386 processor. In order to provide this power, 167 million transistors are used. Needless to say, designing a circuit of this size and making sure that it works properly are extremely challenging tasks.

No matter how fast and powerful a circuit is, it may become useless if its behavior differs from what is expected. To ensure the functional correctness of a circuit, tremendous resources have been devoted to verification. As a result, verification already accounts for two thirds of the circuit design cycle and the overall design/verification effort [13, 110]. However, many ICs are still released with latent errors, demonstrating how poor the current techniques are in ensuring functional correctness. To this end, various estimates indicate that functional errors are currently responsible for 40% of failures at the first circuit production [13, 110], and the growth in design size and overall complexity is much faster than the growth of engineers' verification capabilities. Therefore, verification begins to limit the features that can be implemented in a design [49], essentially becoming the bottleneck that hampers the improvement of modern electronic devices.

To address this problem, the current trend is to automate testbench generation and verification in order to find design bugs more thoroughly. Once a bug is found, however, fixing the bug is still mostly manual and ad hoc. Therefore, engineers often need to spend a great amount of time analyzing and fixing the design errors. Although waveform viewers and simulators are great aids to this end, there are currently no good methodologies and algorithms that can automate the debugging process. The lack of automatic debugging methodologies not only slows down the verification process but also makes thorough design verification more difficult. To this end, Intel's latest Core 2 Duo processor can serve as an example [160]: a detailed analysis of published errata performed by Theo de Raadt in mid 2007 identified 20–30 bugs that cannot be masked by changes in Basic Input/Output System (BIOS) and operating systems, while some may be exploited by malicious software. De Raadt estimates that Intel will take a year to fix these bugs in Core 2 processors. It is particularly

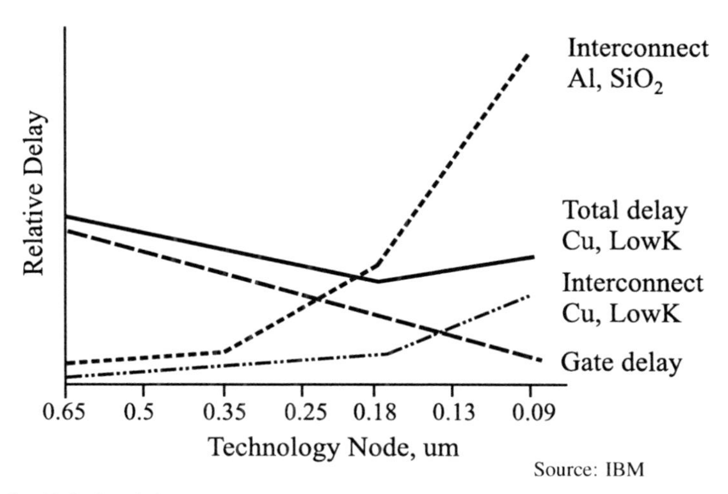

Figure 1.1. Relative delay due to gate and interconnect at different technology nodes. Delay due to interconnect becomes larger than the gate delay at the 90 nm technology node.

alarming that these bugs escaped Intel's verification and validation methodologies, which are considered among the most advanced and effective in the industry. A recent bug found in AMD Phenom processor [152] further shows how difficult the verification problem is.

Another challenge comes from the improvement in IC manufacturing technology that allows smaller transistors to be created on a silicon die. This improvement enables the transistors to switch faster and consume less power. However, the delay due to interconnect is also becoming more significant because of the miniaturization in transistor size. As Figure 1.1 shows, delay due to interconnect already becomes larger than the gate delay at the 90 nm technology node. To mitigate this effect, various physical synthesis techniques and even more powerful optimizations such as retiming are used [120]. These optimizations further exacerbate the verification problem in several ways. First, since *Electronic Design Automation (EDA)* tools may still contain unexpected bugs [9], it is important to verify the functional correctness of the optimized circuit. However, once a bug is found, it is very difficult to pinpoint the optimization step that caused the bug because a large number of circuit modifications may have been performed, which makes repairing the error very challenging. Second, to preserve the invested physical synthesis effort, bugs found in late design stages must be repaired carefully so as to preserve previous optimization effort. This is significantly different from traditional design approaches, which restrict bug-fixing to the original high-level description of the circuit and resynthesize it from scratch after every such fix. In summary, the increase in circuit complexity and miniaturization in transistor size make verification and debugging much more difficult than they were just ten years ago.

To support the miniaturization of CMOS circuits, the required masks also become much more sophisticated and expensive. As Figure 1.2 shows [153], mask cost already approaches 3 million dollars per set at the 65 nm technology node. This cost makes any functional mistakes after circuit production very expensive to fix, not to mention the loss in revenue caused by delayed market entry may be even higher than the mask cost. In addition, due to the lack of automatic post-silicon debugging methodologies, repairing design errors post-silicon is much more challenging than repairing them pre-silicon. As a result, it is important to detect and repair design errors as early in the circuit design flow as possible. On the other hand, any post-silicon error-repair technique that allows the reuse of lithography masks can also alleviate this problem.

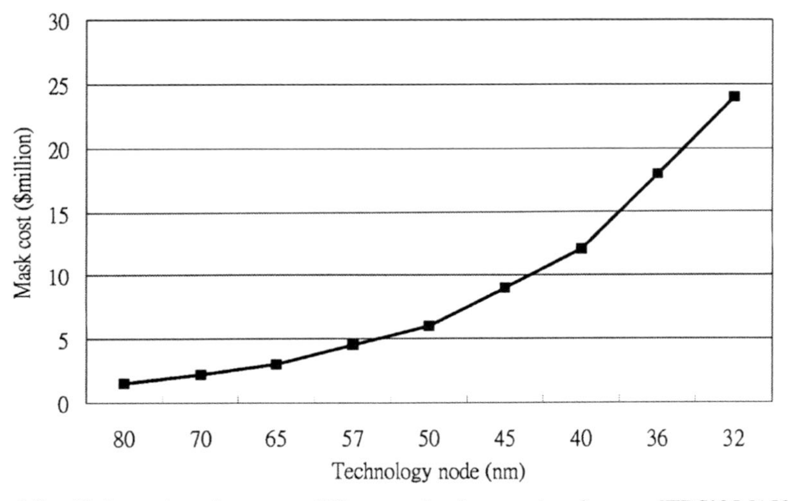

Figure 1.2. Estimated mask costs at different technology nodes. Source: ITRS'05 [153].

1.2 State of the Art

To ensure the functional correctness of a circuit, the current trend is to improve its verification. Among the techniques and methodologies available for functional verification, simulation-based verification is prevalent in industry because of its linear and predictable complexity and its flexibility to be applied, in some form, to any design. The simplest verification method, called *direct test*, is to manually develop suites of input stimuli to test the circuit. Since developing the test suites can be tedious and time consuming, a more flexible methodology called *random simulation* is often used. Random simulation involves connecting a logic simulator with stimuli coming from a constraint-based random generator, that is, an engine that can automatically produce random legal inputs for the design at a very high rate, based on a set of rules (or constraints) derived from the specification document. In order to detect bugs, assertion statements, or checkers, are embedded in the design and continuously monitor the simulated activity for anomalies. When a bug is detected, the simulation trace leading to it is stored and can be replayed later to analyze the conditions that led to the failure. This trace is called a *bug trace*.

Although simulation is scalable and easy to use, it cannot guarantee the correctness of a circuit unless all possible test vectors can be exhaustively tried. Therefore, another verification approach called *formal verification* began to attract increasing attention from industry. Formal verification tools use mathematical methods to prove or disprove the correctness of a design with respect to a certain formal specification or property. In this way, complete verification can be achieved. For example, *symbolic simulation, Bounded Model Checking (BMC)* and *reachability analysis* [16, 72] all belong to this genre. However, formally verifying the correctness of a design tends to become more difficult when design gets larger. Therefore, currently it is often applied to small and critical components within large designs only.

To leverage the advantages of both simulation and formal approaches, a hybrid verification methodology, called *semi-formal verification*, has recently become more popular [70]. Semi-formal techniques strive to provide better scalability with minimal loss in their verification power. To achieve these goals, semi-formal techniques often use heuristics to intelligently select the verification methods to apply, either simulation or formal methods. When the current methods run out of steam, they switch to other methods and continue verification based on previous results. In this way, semi-formal techniques are able to provide a good balance between scalability and verification power.

The verification techniques described so far focus on detecting design errors. After errors are found, the causes of the errors must be identified so that the errors can be corrected. Automatic error diagnosis and correction at the gate level have been studied for decades because this is the level at which the circuits were traditionally designed. To simplify error diagnosis and correction, Abadir et al. [1] proposed an error model to capture the bugs that occur frequently, which has been used in many subsequent studies [86, 130]. While early work in this domain often relies on heuristics and special error models [1, 54, 86, 97, 130], recent improvements in error-repair theories and Boolean-manipulation technologies have allowed more robust techniques to be developed [5, 6, 125, 117, 126, 142]. These techniques are not limited by specific error models and have more comprehensive error diagnosis or correction power than previous solutions.

After automatic logic-synthesis tools became widely available, design tasks shifted from developing gate-level netlists to describing the circuit's functions at a higher-level abstraction, called the *Register-Transfer Level (RTL)*. RTL provides a software-like abstraction that allows designers to concentrate on the functions of the circuit instead of its detailed implementations. Due to this abstraction, gate-level error diagnosis and correction techniques cannot be applied to the RTL easily. However, this is problematic because most design activity takes place at the RTL nowadays. To address this problem, Shi et al. [122] and Rau et al. [111] employed a software-analysis approach to identify statements in the RTL code that may be responsible for the design errors. However, these techniques can return large numbers of potentially erroneous sites. To narrow down the errors, Jiang et al. [75] proposed a metric to prioritize the errors. Although their techniques can facilitate error diagnosis, error correction remains manual. Another approach proposed by Bloem et al. [18] formally analyzes the RTL code and the failed properties, and it is able to diagnose and repair design errors. However, their approach is not scalable due to the heavy use of formal-analysis methods. Since more comprehensive RTL debugging methodologies are still currently unavailable, automatic RTL error repair remains a difficult problem and requires more research.

Another domain that began to attract people's attention is that of post-silicon debugging. Due to the unparalleled complexity of modern circuits, more and more bugs escaped pre-silicon verification and were found post-silicon. Post-silicon debugging is considerably more difficult than pre-silicon debugging due to its limited observability: without special constructs, only signals at the primary inputs and outputs can be observed. Even if the bug can be diagnosed and a fix is found, changing the circuit on a silicon die to verify the fix is also difficult if at all possible. To address the first problem, *scan chains* [23] have been used to observe the values in registers. To address the second problem, *Focused Ion Beam (FIB)* has been introduced to physically change the metal connections between transistors on a silicon die. Alternatively, techniques that use programmable logic have been proposed [93] for this purpose. A recent start-up company called DAFCA [159] proposed a more comprehensive approach that addresses both problems by inserting special constructs before the circuit is taped out. Although these techniques can facilitate post-silicon debugging, the debugging process itself remains manual and ad hoc. Therefore, post-silicon debugging is still mostly an art, not a science [67].

1.3 New Opportunities

Despite the vast amount of verification and debugging effort invested in modern circuits, these circuits are still often released with latent bugs, showing the deficiency of current methodologies. One major reason is that existing error diagnosis and correction techniques typically lack the power and scalability to handle the complexity of today's designs. Another reason is that existing verification techniques often focus on finding design errors without considering how the errors should be fixed. Therefore, the bug traces produced by verification can be prohibitively long, making human analysis extremely difficult and further hampering the deployment of automatic error-repair tools. As a result, error repair remains a demanding, semi-manual process that often introduces new errors and consumes valuable resources, essentially undermining thorough verification.

To address these problems, we propose a framework called *FogClear* that automates the error-repair processes at various design stages, including front-end design, back-end logic design, back-end physical design and post-silicon debugging. We observe that major weakness exists in several key components required by automatic error repair, and this deficiency may limit the power and scalability of the framework. To ensure the success of our methodologies, we also develop innovative data structures, theories and algorithms to strengthen these components. Our enhanced components are briefly described below.

- *Butramin* reduces the complexity of bug traces produced by verification for easier error diagnosis.

- *REDIR* utilizes bug traces to automatically correct design errors at the RTL.

- *CoRé* utilizes bug traces to automatically correct design errors at the gate level.

- *InVerS* monitors physical synthesis optimizations to identify potential errors and facilitates debugging.

- To repair post-silicon electrical errors, we propose *SymWire*, a symmetry-based rewiring technique, to perturb the layout and change the electrical characteristics of the erroneous wires. In addition, we devise a *SafeResynth* technique to identify alternative signal sources that can generate the same signal, and use the identified sources to change the wiring topology in order to repair electrical errors.

- To repair post-silicon functional errors, we propose *PAFER* and *PARSyn* that can change a circuit's functionality via wire reconnections. In this way, transistor masks can be reused and respin cost can be reduced.

The strength of these components stems from the intelligent combination of simulation and formal verification techniques. In particular, recent improvements in *SATisfiability (SAT)* solvers provide the power and scalability to handle modern circuits. By enhancing the power of key components, as well as unifying verification and debugging into the same framework, the FogClear framework promises to facilitate the debugging processes at various design stages, thus improving the quality of electronic devices in several categories.

1.4 Key Innovations and Book Outline

In this book we present advanced theories and methodologies that address the error diagnosis and correction problem of digital circuits. In addition, we propose scalable and powerful algorithms to match the error-repair requirements at different design stages. On the methodological front, we promote interoperability between verification and debugging by devising new design flows that automate the error-repair processes in front-end design, back-end logic design, back-end physical design and post-silicon debugging. On the theoretical front, we propose a counterexample-guided error-repair framework that performs abstraction using signatures, which is refined by counterexamples that fail further verification. This framework integrates verification into debugging and scales much further than existing solutions due to its innovative abstraction mechanism. To support the error-correction needs in the framework, we design two resynthesis algorithms, which are based on a compact encoding of resynthesis information called *Pairs of Bits to be Distinguished (PBDs)*. These resynthesis techniques allow us to repair design errors effectively. We also develop a comprehensive functional symmetry detector that can identify permutational, phase-shift, higher-level, as well as composite input and output symmetries. We apply this symmetry-detection technique to rewiring and use it to repair post-silicon electrical errors.

To enhance the robustness and power of FogClear, it is important to make sure that each component used in the framework is scalable and effective. We observe that existing solutions exhibit major weakness when we implement several components critical to our framework. Therefore, we develop new techniques to strengthen these components. In particular, we observe that verification tools often strive to find many errors without considering how these errors should be resolved. As a result, the returned bug traces can be tremendously long. Existing solutions to reduce the complexity of the traces, however, rely heavily on formal methods and are not scalable [53, 65, 68, 77, 113, 115, 121]. To this end, we propose a bug trace minimizer called Butramin using several simulation-based methods. This minimizer scales much further than existing solutions and can handle more realistic designs. Another component that receives little attention is RTL error diagnosis and correction. Although techniques that address this problem began to emerge in the past few years [18, 75, 111, 122, 126], they are not accurate or scalable enough to handle today's circuits. To design an effective automatic RTL debugger, we extend state-of-the-art gate-level solutions to the RTL. Our empirical evaluation shows that our debugger is powerful and accurate, yet it manages to avoid drawbacks common in gate-level error analysis and is highly scalable. On the other end of the design flow, we observe that post-silicon debugging is often ad hoc and manual. To solve this problem, we propose the concept of physical safeness to identify physical synthesis techniques that are suitable for this design stage. In addition, we propose several new algorithms that can repair both functional and electrical errors on a silicon die.

The rest of the book is organized as follows. Part I, which includes Chapters 2 and 3, provides necessary background and illustrates prior art. In particular, Chapter 2 outlines the current design and verification landscapes. In this chapter, we discuss the front-end design flow, followed by back-end design flows and the post-silicon debugging process. Chapter 3 introduces several traditional techniques for finding and fixing bugs, including simulation-based verification, formal-verification methods, design-for-debug constructs and post-silicon metal fix.

Part II, which includes Chapters 4, 5, 6, and 7, illustrates our FogClear methodologies and presents our theoretical advances in error repair. We start from the proposed FogClear design and verification methodologies in Chapter 4. In this chapter, we describe how our methodologies address the error-repair problems at different design stages. Chapter 5 then illustrates our gate-level functional error correction framework, *CoRé*, that uses counterexamples reported by verification to automatically repair design errors at the gate level [40, 41]. It scales further than existing techniques due to its intelligent use of signature-based abstraction and refinement. To support the error-correction requirements in CoRé, we propose two innovative resynthesis techniques, *Distinguishing-Power Search (DPS)* and *Goal-Directed Search (GDS)* [40, 41], in Chapter 6. These techniques can be used to find resynthesized netlists that change the functionality of the circuit to match a given specification. To allow efficient manipulation of logic for resynthesis, we also describe a compact encoding of required resynthesis information in the chapter, called *Pairs of Bits to be Distinguished (PBDs)*. Finally, Chapter 7 presents our comprehensive symmetry-detection algorithm based on graph-automorphism, and we applied the detected symmetries to rewiring in order to optimize wirelength [36, 37]. This rewiring technique is also used to repair electrical errors as shown in Section 11.4.1.

Part III, which includes Chapters 8, 9, 10, 11, and 12, discusses specific FogClear components that are vital to the effectiveness of our methodologies. We start from our proposed bug trace minimization technique, *Butramin* [34, 35], in Chapter 8. Butramin considers a bug trace produced by a random simulator or semi-formal verification software and generates an equivalent trace of shorter length. By reducing the complexity of the bug trace, error diagnosis will become much easier. Next, we observe that functional mistakes contribute to a large portion of design errors, especially at the RTL and the gate level. Our solutions to this end are discussed in Chapter 9, which includes gate-level error repair for sequential circuits and RTL error repair [45]. Our techniques can diagnose and repair errors at these design stages, thus greatly saving engineers' time and effort. Since interconnect begins to dominate delay and power consumption at the latest technology nodes, more aggressive physical synthesis techniques are used, which exacerbates the already difficult verification problem. In Chapter 10 we describe an incremental verification framework, called *InVerS*, that can identify potentially erroneous netlist transformations produced by physical synthesis [44]. InVerS allows early detection of bugs and promises to reduce the debugging effort.

After a design has been taped-out, bugs may be found on a silicon die. We notice that due to the special physical constraints in post-silicon debugging, most existing pre-silicon error-repair techniques cannot be applied to this design stage. In Chapter 11 we first propose the concept of physical safeness to measure the impact of physical optimizations on the layout [38, 39], and then use it to identify physical synthesis techniques that can be applied post-silicon. To this end, we observe that safe techniques are particularly suitable for post-silicon debugging; therefore, we propose a *SafeResynth* technique based on simulation and on-line verification. We then illustrate how functional errors can be repaired by our *PAFER* framework and *PARSyn* algorithm [42, 43]. In addition, we describe how to adapt symmetry-based rewiring and SafeResynth for electrical error repair. In Chapter 12 we describe new methodologies for spare-cell insertion, which are important to the success of post-silicon debugging [46]. Finally, Chapter 13 concludes this book by providing a summary of key techniques described in this book.

Chapter 2

CURRENT LANDSCAPE IN DESIGN AND VERIFICATION

Before delving into error-repair techniques, we are going to review how digital circuits are developed and verified first. In this chapter we describe current flows for front-end design, back-end logic design, back-end physical design and post-silicon debugging. We also discuss the bugs that may appear at each design stage, as well as the current verification and debugging methodologies that attack them.

2.1 Front-End Design

Figure 2.1 illustrates the current front-end design flow. Given a specification, typically three groups of engineers will work on the same design, including architecture design, testbench creation and RTL development[1]. The flow shown in Figure 2.1 uses simulation-based verification; however, flows using formal verification are similar. Chapter 3 provides more detailed discussions on these verification methods.

In this design flow, the architecture group first designs a high-level initial model using high-level languages such as C, C++, SystemC, Vera [163], e [150] or SystemVerilog. At the same time, the verification group develops a testbench to verify the initial model. If verification fails, the testbench and/or model need to be corrected, after which their correctness is verified again. This process keeps repeating until the high-level model passes verification. At this time, a golden high-level model and testbench will be produced. They will be used to verify the RTL initial model developed by the RTL group. If verification passes, an RTL golden model will be produced. If verification

[1]Although there may be other groups of engineers working on other design aspects, such as power, we do not consider them in this design flow.

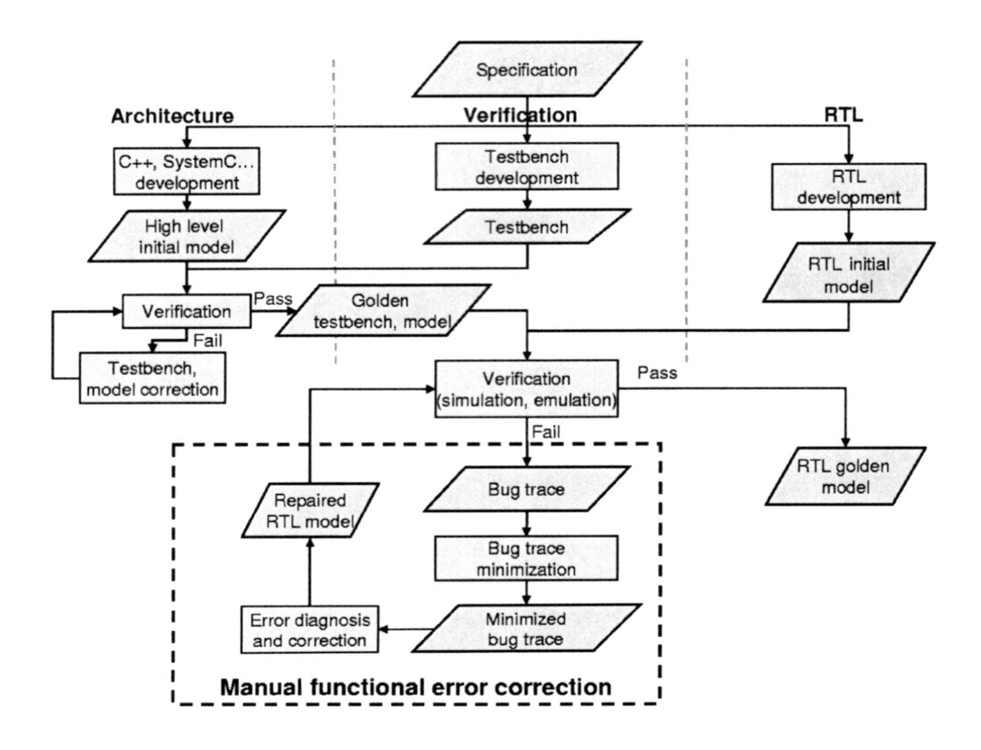

Figure 2.1. The current front-end design flow.

fails, the RTL model contains bugs and must be fixed. Usually, a bug trace that exposes the bugs in the RTL model will be returned by the verification tool.

To address the debugging problem, existing error-repair techniques often partition the problem into two steps. In the first step, the circuit is diagnosed to identify potential changes that can alter the incorrect output responses. In the second step, the changes are implemented. The first step is called *error diagnosis*, and the second step is called *error correction*. Currently, functional error diagnosis and correction are often performed manually using the steps described below. This manual error-repair procedure is also shown in the "Manual functional error correction" block in Figure 2.1.

1 The bug trace is minimized to reduce its complexity for easier error diagnosis.

2 The minimized bug trace is diagnosed to find the cause of the bugs. Debugging expertise and design knowledge are usually required to find the cause of the bugs.

3 After the cause of the bugs is found, the RTL code must be repaired to remove the bugs. The engineer who designed the erroneous block is usually responsible for fixing the bugs.

4 The repaired RTL model needs to be verified again to ensure the correctness of the fix and prevent new bugs from being introduced by the fix.

Errors in semiconductor products have different origins, ranging from poor specifications, miscommunication among designers, and designer's mistakes – conceptual or minor. Table 2.1 lists 15 most common error categories in microprocessor designs specified at the Register-Transfer Level (RTL), collected from student projects at the University of Michigan between 1996 and 1997 [27]. In addition to academic data, Intel also analyzed the sources of bugs found in the Pentium 4 processor [12], and the major categories are as follows: RTL coding (18.1%), microarchitecture (25.1%), logic/microcode changes (18.4%) and architecture (2.8%).

Table 2.1. Distribution of design errors (in %) in seven microprocessor projects.

Error category	Microprocessor project							Ave.
	LC2	DLX1	DLX2	DLX3	X86	FPU	FXU	
Wrong signal source	27.3	31.4	25.7	46.2	32.8	23.5	25.7	30.4
Missing instance	0.0	28.6	20.0	23.1	14.8	5.9	15.9	15.5
Missing inversion	0.0	8.6	0.0	0.0	0.0	47.1	16.8	10.3
New category (Sophisticated)	9.1	8.6	0.0	7.7	6.6	11.8	4.4	**6.9**
Unconnected inputs	0.0	8.6	14.3	7.7	8.2	5.9	0.9	6.5
Missing inputs	9.1	8.6	5.7	7.7	11.5	0.0	0.0	6.1
Wrong gate/module type	13.6	0.0	11.4	0.0	9.8	0.0	0.0	5.0
Missing item/factor	9.1	2.9	5.7	0.0	0.0	0.0	4.4	3.2
Wrong constant	9.1	0.0	2.9	0.0	0.0	0.0	9.7	3.1
Always statement	9.1	0.0	2.9	0.0	0.0	0.0	2.7	2.1
Missing latch/flip-flop	0.0	0.0	0.0	0.0	4.9	5.9	0.9	1.7
Wrong bus width	4.5	0.0	0.0	0.0	0.0	0.0	7.1	1.7
Missing state	9.1	0.0	0.0	0.0	0.0	0.0	0.0	1.3
Conflicting outputs	0.0	0.0	0.0	7.7	0.0	0.0	0.0	1.1
Conceptual error	0.0	0.0	2.9	0.0	3.3	0.0	0.9	**1.0**

Reproduced from [27, Table4], where the top 15 most-common errors are shown. "New category" includes timing errors and sophisticated, difficult-to-fix errors.

Since the purpose of RTL development is to describe the logic function of the circuit, the errors occurring at the RTL are mostly functional. We observe from Table 2.1 that most errors are simple in that they only require the change of a few lines of code, while complex errors only contribute to 6.9% of the total errors. This is not surprising since competent designers should be able to write code that is close to the correct one [60]. However, finding and fixing those bugs are still challenging and time-consuming. Since fixing errors at later design stages will be much more difficult and expensive, it is especially important to make sure that the RTL code describes the function of the circuit correctly.

To address this problem, techniques that focus on RTL debugging have been developed recently. The first group of techniques [111, 122] employ a software-analysis approach that implicitly uses multiplexers (MUXes) to identify statements in the RTL code that are responsible for the errors. However, these techniques can return large numbers of potentially erroneous sites. To address this problem, Jiang et al. [75] proposed a metric to prioritize the errors. Their techniques greatly improve the quality of error diagnosis, but error correction remains manual. The second group of techniques, such as those in [18], uses formal analysis of an HDL description and failed properties; because of that these techniques can only be deployed in a formal verification framework, and cannot be applied in a simulation-based verification flow common in the industry today. In addition, these techniques cannot repair identified errors automatically. Finally, the work by Staber *et al.* [126] can diagnose and correct RTL design errors automatically, but it relies on state-transition analysis and hence, it does not scale beyond tens of state bits. In addition, this algorithm requires a correct formal specification of the design, which is rarely available in today's design environments because its development is often as challenging as the design process itself. In contrast, the most common type of specification available is a high-level model, often written in a high-level language, which produces the correct I/O behavior of the system. As we show in Section 4.1, our FogClear methodology is scalable and can automate both error diagnosis and repair at the RTL. In addition, it only requires the correct I/O behavior to be known.

2.2 Back-End Logic Design

Front-end design flow produces an RTL model that should be functionally correct. The next step is to produce a gate-level netlist that has the same functionality by performing back-end logic design, followed by back-end physical design that generates the layout. This section discusses the logic design flow, and the next section describes the physical design flow.

Figure 2.2 shows the current back-end logic design flow. Given an RTL golden model, this flow produces a gate-level netlist that efficiently implements the logic functions of the RTL model. This goal is achieved by performing logic synthesis and various optimizations, which are already highly automated. However, since logic synthesis may not capture all the behavior of the RTL

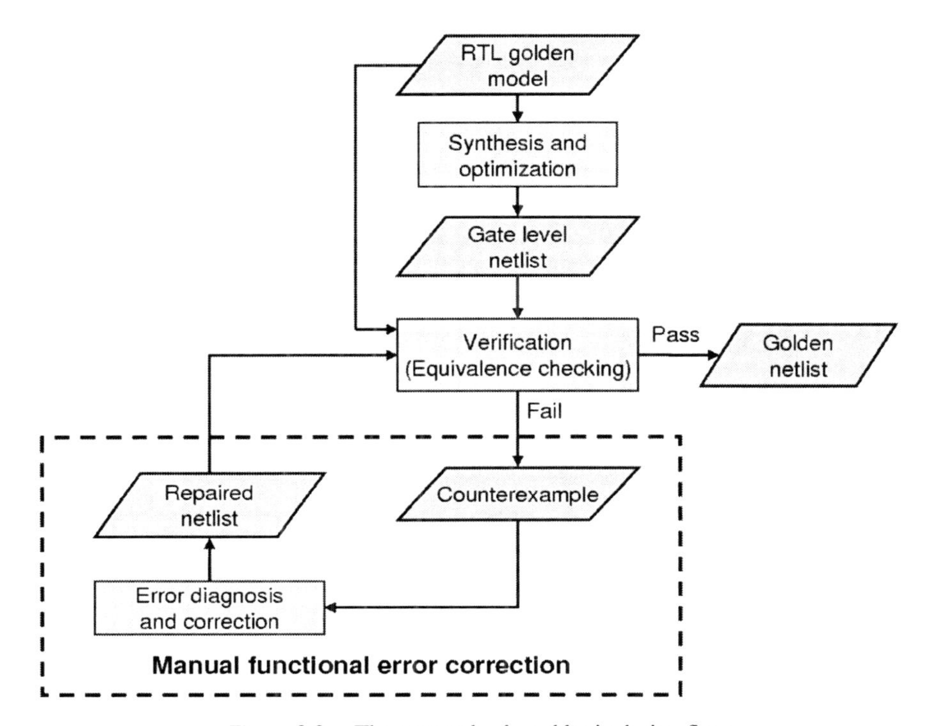

Figure 2.2. The current back-end logic design flow.

code faithfully [19], it is possible that the produced netlist does not match the RTL model. In addition, unexpected bugs may still exist in synthesis tools [9]. Therefore, verification is still required to ensure the correctness of the netlist.

Another reason to fix functional errors at the gate-level instead of the RTL is to preserve previous design effort, which is especially important when the errors are discovered at a late stage of the design flow. In the current design methodology, functional errors discovered at an early stage of the design flow are often conceivable to be fixed by changing the RTL code and synthesizing the entire netlist from scratch. However, such a design strategy is typically inefficient when the errors are discovered at a late stage of the design flow because previously performed optimizations will be invalidated. Additionally, gate-level bug-fixing offers possibilities not available when working with higher-level specifications, such as reconnecting individual wires, changing individual gate types, etc.

One way to verify the correctness of the netlist is to rerun the testbench developed for the RTL model, while Figure 2.2 shows another approach where the netlist is verified against the RTL model using equivalence checking. In either approach, when verification fails, a counterexample (or a bug trace for the simulation-based approach) will be produced to expose the mismatch. This counterexample will be used to debug the netlist.

Before logic synthesis was automated, engineers designed digital circuits at the gate level or had to perform synthesis themselves. In this context, Abadir et al. [1] proposed an error model to capture the bugs that occur frequently at this level. In the current design methodology, however, gate-level netlists are often generated via synthesis tools. As a result, many bugs that exist in a netlist are caused by the erroneous RTL code and may not be captured by this model. On the other hand, bugs introduced by *Engineering Change Order (ECO)* modifications or EDA tools can often be categorized into the errors in this model.

Similar to RTL debugging, existing gate-level error-repair techniques also partition the debugging problem into *Error Diagnosis (ED)* and *Error Correction (EC)*. The solutions that address these two problems are described below.

Error diagnosis has been extensively studied in the past few decades. For example, early work by Madre et al. [97] used symbolic simulation and Boolean equation solving to identify error locations, while Kuo [86] used *Automatic Test Pattern Generation (ATPG)* and don't-care propagation. Both of these techniques are limited to single errors only. Recently, Smith et al. [125] and Ali et al. [6] used Boolean satisfiability (SAT) to diagnose design errors. Their techniques can handle multiple errors and are not limited to specific error models. We adopt these techniques for error diagnosis because of their flexibility, which will be described in detail in Chapter 5. To further improve the scalability of SAT-based error diagnosis, Safarpour et al. [116] proposed an abstraction-refinement scheme for sequential circuits by replacing registers with primary inputs, while Ali et al. [5] proposed the use of *Quantified Boolean Formulas (QBF)* for combinational circuits.

Error correction implements new logic functions found by diagnosis to fix the erroneous behavior of the circuit. Madre et al. [97] pointed out that the search space of this problem is exponential and, in the worst case, is similar to that of synthesis. As a result, heuristics have been used in most publications. Chung et al. [54] proposed a *Single-Logic-Design-Error (SLDE)* model in their *ACCORD* system, and were able to detect and correct errors that comply with the model. To further reduce the search space, they also proposed *screen tests* to prune the search. The *AutoFix* system from Huang et al. [73] assumed that the specification is given as a netlist and equivalence points can be identified between the specification and the circuit. The error region in the circuit can then be reduced by replacing the equivalent points with pseudo-primary inputs and outputs, and the errors are corrected by resynthesizing the new functions using the pseudo-primary inputs. Lin et al. [92] first synthesized and minimized the candidate functions using BDDs, and then replaced the inputs to the BDDs by signals in the circuit to reuse the existing netlist. Swamy et al.

[128] synthesized the required functions by using the signals in *minimal regions*. Work by Veneris et al. [130] handled this problem by trying possible fixes according to the error model proposed by Abadir et al. [1]. Staber et al. [126] proposed a theoretically sound approach that fixes design errors by preventing the reach of bug states, which can also be applied to RTL debugging and software error correction. Although these techniques have been successful to some degree, their correction power is often limited by the heuristics employed or the logic representations used. For example, either the error must comply with a specific error model [54, 130] or the specification must be given [54, 73, 126]. Although the work by Lin et al. [92] and Swamy et al. [128] has fewer restrictions, their techniques require longer runtime and do not scale well due to the use of BDDs. The work by Staber et al. [126] also does not scale well because of the heavy use of state-transition analysis. A recent approach proposed by Yang et al. [142] managed to avoid most drawbacks in current solutions. However, their techniques are based on *Sets of Pairs of Functions to be Distinguished (SPFDs)*, which are more difficult to calculate and represent than the signature-based solutions shown in Section 6.1.2.

Table 2.2. A comparison of gate-level error diagnosis and correction techniques.

Technique	ED/ EC	No. of errors	Error model	Scalability	Requirement
ACCORD [54]	Both	Single	SLDE	Moderate (BDDs)	Functional specification
AutoFix [73]	Both	Multiple	None	Moderate (BDDs)	Golden netlist
Kuo [86]	ED	Single	Abadir	Good (ATPG)	Test vectors
Lin [92]	Both	Multiple	None	Moderate (BDDs)	Golden netlist
Madre [97]	Both	Single	PRIAM	Moderate	Functional specification
Smith [125]	ED	Multiple	None	Good (SAT)	Test vectors
Staber [126]	Both	Multiple	None	Moderate (State analysis)	Functional specification
Veneris [130]	Both	Multiple	Abadir	Good (ATPG)	Test vectors
CoRé (Chapter 5)	Both	Multiple	None	Good (SAT, signatures)	Test vectors

A comparison of the work presented in this book (CoRé) with previous error diagnosis and correction techniques is given in Table 2.2. In the table, "No.

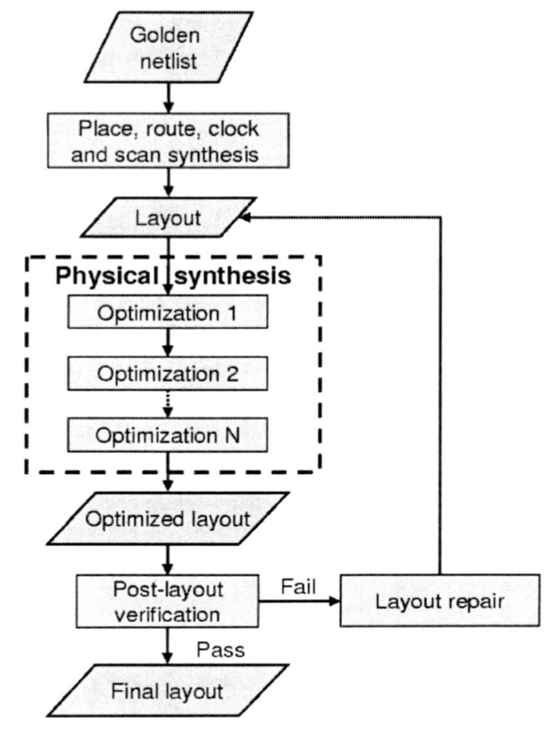

Figure 2.3. The current back-end physical design flow.

of errors" is the number of errors that can be detected or corrected by the technique. Our gate-level error-repair framework, CoRé, will be described in detail in Chapter 5.

2.3 Back-End Physical Design

The current back-end physical design flow is shown in Figure 2.3. Starting from the golden netlist, place and route is first performed to produce the layout. Sometimes clock or scan synthesis also needs to be performed, as well as physical synthesis that optimizes timing or power of the circuit. Post-layout verification is then carried out to ensure the correctness of the layout. If verification fails, the cause of the error must be diagnosed. If the error is due to timing violations, layout timing repair needs to be performed to fix the error, usually via more iterations of physical synthesis. Since bugs are still common in today's logic and physical synthesis tools [9], logic errors may still be introduced. When this happens, the manual functional error correction process shown in Figure 2.2 needs to be performed, and this process will produce a repaired netlist. The layout is then modified to reflect the change in the netlist, after which its correctness is verified again.

2.4 Post-Silicon Debugging

Figure 2.4 shows the current post-silicon debugging flow. To verify the correctness of a silicon die, engineers apply numerous test vectors to the die and then check their output responses. If the responses are correct for all the applied test vectors, then the die passes verification. If not, then the test vectors that expose the design errors become the bug trace that can be used to diagnose and correct the errors. The trace will then be diagnosed to identify the root causes of the errors. Typically, there are three types of errors: functional, electrical and manufacturing/yield. In this work we only focus on the first two types.

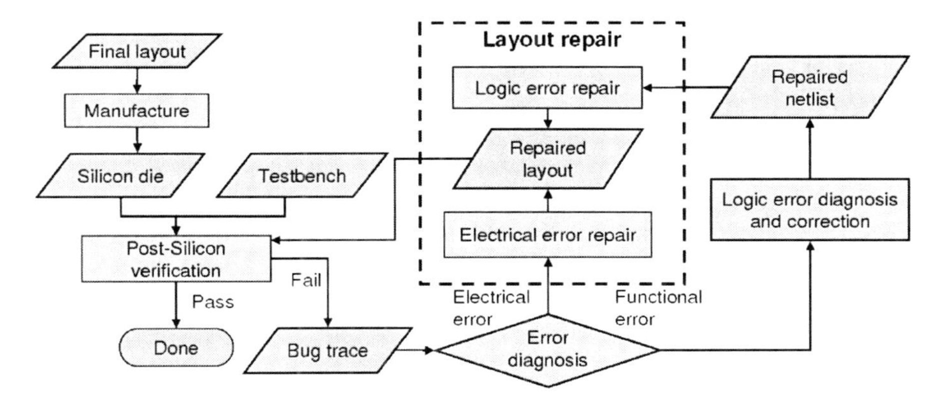

Figure 2.4. The current post-silicon debugging flow.

After errors are diagnosed, the layout is modified to correct them, and the repaired layout must be verified again. This process is repeated until no more errors are exposed. In post-silicon debugging, however, it is often unnecessary to fix all the errors because repairing a fraction of the errors may be sufficient to enable further verification. For example, a processor may contain a bug in its ALU and another one in its branch predictor. If fixing the bug in the ALU is sufficient to enable further testing, then the fix in the branch predictor can be postponed to the next respin. On the other hand, all the bugs should be fixed before the chip is shipped to the customers.

Josephson documented the major silicon failure mechanisms in microprocessors [78], where the most common failures (excluding dynamic logic) are drive strength (9%), logic errors (9%), race conditions (8%), unexpected capacitive coupling (7%), and drive fights (7%). Another important problem at the latest technology nodes are antenna effects, which can damage a circuit during its manufacturing or reduce its reliability. These problems often can only be identified in post-silicon debugging.

Pre-silicon and post-silicon debugging differ in several significant ways. First, conceptual bugs that require deep understanding of the chip's functionality are predominantly introduced when the chip is being designed and well before the first silicon is available, and such bugs may not be fixable by automatic tools. As Table 2.1 shows, however, complex and conceptual errors only contribute to 7.9% of the errors at early design stages, and such errors can often be caught by pre-silicon verification. As a result, post-silicon functional bugs are mostly subtle errors that only affect the output responses of a few input vectors, and their fixes can usually be implemented with very few gates. As an analysis of commercial microprocessors suggests [133], faults in control logic contribute to 52% of the total errors, which are typically subtle and only appear in rare corner-cases. However, repairing such errors requires the analysis of detailed layout information, making it a highly tedious and error-prone task. As we show in Chapter 11, our work can automate this process. Second, errors found post-silicon typically include functional and electrical problems, as well as those related to manufacturability and yield. However, issues identified pre-silicon are predominantly related to functional and timing errors.[2] Problems that manage to evade pre-silicon validation are often difficult to simulate, analyze and even duplicate. Third, the observability of the internal signals on a silicon die is extremely limited. Most internal signals cannot be directly observed, even in designs with *built-in scan chains* (see Section 3.3.1) that enable access to sequential elements. Fourth, verifying the correctness of a fix is challenging because it is difficult to physically implement a fix in a chip that has already been manufactured. Although techniques such as FIB exist (see Section 3.3.2), they typically can only change metal layers of the chip and cannot create any new transistor (this process is often called *metal fix*). Finally, it is especially important to minimize the layout area affected by each change in post-silicon debugging because smaller changes are easier to implement with good FIB techniques, and there is a smaller risk of unexpected side effects. Due to these unusual circumstances and constraints, most debugging techniques prevalent in early design stages cannot be applied to post-silicon debugging. In particular, conventional physical synthesis and ECO techniques affect too many cells or wire segments to be useful in post-silicon debugging. As illustrated in Figure 2.5(b), a small modification in the layout that sizes up a gate requires changes in all transistor masks and refabrication of the chip. On the other hand, our techniques are aware of the physical constraints and can repair errors with minimal physical changes, as shown in Figure 2.5(c).

[2] Post-silicon timing violations are often caused by electrical problems and are only their symptoms.

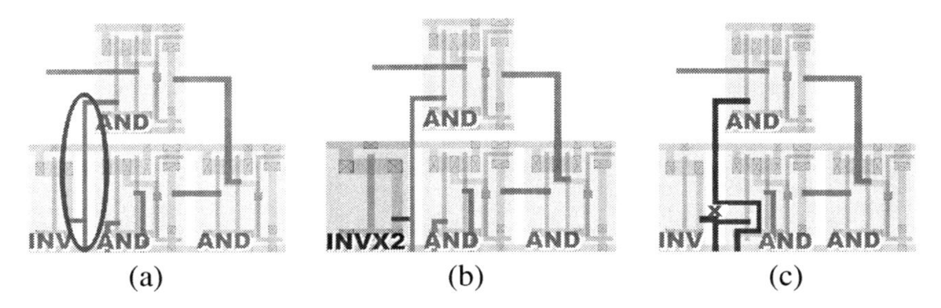

Figure 2.5. Post-silicon error-repair example. (**a**) The original buggy layout with a weak driver (INV). (**b**) A traditional resynthesis technique finds a "simple" fix that sizes up the driving gate, but it requires expensive remanufacturing of the silicon die to change the transistors. (**c**) Our physically-aware techniques find a more "complex" fix using symmetry-based rewiring, and the fix can be implemented simply with a metal fix and has smaller physical impact.

To repair post-silicon functional errors, the current trend is to provide more visibility and controllability of the silicon die. For example, most modern designs incorporate a technique, called *scan test* [23], into their chips. This technique allows engineers to observe the values of internal registers and can greatly improve the design signals' observability. In order to change the logic on a silicon die, spare cells are often scattered throughout a design to enable metal fix [79]. The number of spare cells depends on the methodology, as well as the expectation for respins and future steppings, and this number can reach 1% of all cells in mass-produced microprocessor designs. Our methodology to insert spare cells is discussed in Chapter 12. Alternatively, Lin et al. [93] proposed the use of programmable logic for this purpose. DAFCA provides a more comprehensive solution that further improves the observability of silicon dies and enables logic changes on the dies [2, 159]. A success story can be found in [76].

Debugging electrical errors is often more challenging than debugging functional errors because it does not allow the deployment of logic debugging tools that designers are familiar with. In addition, there are various reasons for electrical errors [78], and analyzing them requires profound design and physical knowledge. Although techniques to debug electrical errors exist (e.g., *voltage-frequency Shmoo plots* [10]), they are often heuristic in nature and require abundant expertise and experience. As a result, post-silicon debugging is currently an art, not a science. Even if the causes of the errors can be identified, finding valid fixes is still challenging because most existing resynthesis techniques require changes in transistor cells and do not allow metal fix. To address this problem, techniques that allow post-silicon metal fix have been developed recently, such as ECO routing [140]. However, ECO routing can only repair a fraction of electrical errors because it cannot find layout transformations involving logic changes. To repair more difficult bugs, transformations that also utilize logic information are required. For example, one way to repair a driving strength error is to identify alternative signal sources that also generate the same signal, and this can only be achieved by considering logic information. All these issues will be addressed and solved by the FogClear post-silicon debugging methodology that we present in Chapter 11.

Chapter 3

FINDING BUGS AND REPAIRING CIRCUITS

In the previous chapter we described the current design and verification methodologies at different design stages. In this chapter we take a closer look at the verification techniques used in these methodologies. Among the techniques available for functional verification, simulation-based verification is prevalent in the industry because of its linear and predictable complexity as well as its flexibility in being applied, in some form, to any design. However, simulation can only find bugs that can be exposed by the given stimuli. Therefore, unless all possible input scenarios can be covered by the stimuli, the correctness of the design cannot be guaranteed. To address this problem, formal methods have been developed to prove the correctness of the design under certain pre-defined properties. Nonetheless, their scalability is often limited because the proving process can be very complicated. In addition, developing properties may be as difficult as developing the design itself. Therefore, formal techniques are applied to only a small portion of the current designs. To overcome this problem, hybrid techniques that utilize both simulation and formal methods have been proposed. In this chapter, we first review simulation-based verification techniques. Next, we describe commonly used formal methods. Finally, we introduce the scan chain Design-for-Debugging (DFD) construct and the metal fix technique that facilitate post-silicon debugging.

3.1 Simulation-Based Verification

Simulation is the most commonly used technique for verifying the correctness of a design. In its simplest form, called *direct test*, engineers manually develop the test vectors that are applied to the design and then inspect their output responses. Developing the test suites, however, can be costly and time-consuming. In addition, scenarios not considered by the designers may be overlooked by the test developers as well. Therefore, techniques that automate testbench generation have been proposed to avoid the bias from human engineers. A common methodology to this context is *constrained-random simulation*. It involves connecting a logic simulator with stimuli coming from a constraint-based random generator, i.e., an engine that can automatically produce random legal inputs for the design at a very high rate based on a set of rules (or constraints) derived from the specification.

A fast simulator is the core of simulation-based verification methodologies; therefore, in this section we review two commonly used simulation algorithms. Since the quality of test vectors generated in constrained-random simulation greatly affects the thoroughness of verification, we also review several solutions that improve this quality.

3.1.1 Logic Simulation Algorithms

Logic simulation mimics the operation of a digital circuit by calculating the outputs of the circuit using given input stimuli. For example, if a 0 is applied to the input of an inverter, logic simulation will produce a 1 on its output. Algorithms that perform simulation can be categorized into two major types: oblivious and event-driven [11]. In the oblivious algorithm, all gates are simulated at each time point. In event-driven simulation, value changes in the netlist are recorded, and only the gates that might cause further changes are simulated. Event-driven algorithms are potentially more efficient than oblivious algorithms because they only simulate the part of the netlist that had their values changed; however, the overhead to keep track of the gates that should be simulated is also a concern.

A typical oblivious simulation algorithm works as follows:

1 A linear list of gates is produced by levelizing the netlist. Gates closer to the primary inputs (i.e., those at lower levels of logic) are placed on the front of the list.

2 At each time point, all the gates in the list are simulated. Since gates with smaller levels of logic are simulated first, the simulation values at the inputs of the gate currently being simulated are always valid. As a result, the simulation value at the gate's output is also correct.

Event-driven algorithms are more complicated than oblivious ones because the algorithms must keep track of the gates that need to be resimulated. One such algorithm, proposed by Lewis [90], is shown in Figure 3.1. Two phases are used in Lewis' algorithm, including the node phase (also called the event phase) and the gate phase (also called the evaluation phase).

The node phase corresponds to the code labeled "fanout:", and the gate phase corresponds to the code labeled "simulate:". There are two lists that represent the state of the netlist: the first one is for the active nodes, while the

```
1   fanout:
2   foreach node ∈ active_nodes
3       node.val= node.next_val;
4       active_gates = active_gates ∪ node's fanout gates;
5   active_nodes.clear();
6   simulate:
7   foreach gate ∈ active_gates
8       simulate gate;
9       foreach node ∈ gate's output
10          if (node.val != node.next_val)
11                  active_nodes=active_nodes ∪ node;
12  active_gates.clear();
```

Figure 3.1. Lewis' event-driven simulation algorithm.

other one is for the active gates. At each time point, nodes in $active\_nodes$ list are scanned and their fanout gates are added to the $active\_gates$ list. The logic value of each node is also updated from its $next\_val$, and the $active\_nodes$ list is cleared. The $active\_gates$ list is then scanned, and each active gate is simulated. The simulation results will be used to update the $next\_val$ of the gate's output nodes. If the node's new value (in $node.next\_val$) is different from its current value (in $node.val$), the node will be added to the $active\_nodes$ list. The $active\_gates$ list is then cleared. Since gates will be simulated only if their input values change, Lewis' simulation algorithm can avoid redundant computation that simulates gates whose output values will not change.

3.1.2 Improving Test Generation and Verification

One major obstacle in adopting constrained-random simulation into the verification flow is that writing constraints may be difficult: the constraints need to model the environment for the design under verification, and describing the environment using constraints can be challenging. To address this problem, Yuan et al. [144] proposed techniques to generate the constraints using rules specified by designers. In this way, test developers can focus on describing the high-level behavior of the environment and let the tool automatically generate the constraints. Alternatively, commercial *Verification Intellectual Properties (VIPs)* and *Bus Functional Models (BFMs)* are also available for modeling the test environment [147, 151].

The quality of a test suite is determined by the input scenarios that can be explored by its tests. Test vectors that cover corner-case scenarios are often considered as of high quality. Since random simulation tends to cover scenarios that occur frequently, techniques that try to generate tests with higher quality have been proposed. For example, the StressTest technique [132] monitors circuit activities at key signals and uses a Markov-model-driven test generator to cover the corner-case scenarios. Shimizu et al. [124] took another approach by deriving an input generator and a coverage metric from a formal specification first, and then they used the measured coverage to bias the input generator. Recent work by Plaza et al. [108] measures signal activities based on Shannon entropy and uses the measured activities to guide a pattern generator to produce high-quality test vectors.

3.2 Formal Verification

Simulation-based verification uses a large number of input vectors to check a design's responses. Due to the scalability of modern simulators, whole-chip simulation can often be performed. However, it is usually infeasible to simulate all possible input sequences because the number of the sequences is typically large and can even be infinite. As a result, it is difficult to judge whether all possible scenarios have been covered, making complete verification difficult.

Formal verification is a totally different approach. It uses mathematical methods to prove or disprove the correctness of the design with respect to a certain formal specifications or properties. In this way, complete verification can be achieved to the extent described by the specification or properties. However, the complexity of formally verifying a design grows considerably with the size of the circuit, making formal techniques applicable to smaller designs only. As a result, currently it is often used to verify small and critical components within a large design.

In this section we first describe a commonly used problem formulation, the *SATisfiability (SAT)* problem. Next, we briefly introduce several formal verification techniques, including *Bounded Model Checking (BMC)*, *symbolic simulation*, *reachability analysis* and *equivalence checking*.

3.2.1 The Boolean Satisfiability Problem

A Boolean SATisfiability (SAT) problem can be formulated as follows. Given a Boolean expression composed of AND, OR, NOT, variables and parentheses, determine if there is an assignment of true and false values to the variables that makes the expression evaluate to true. If no such assignment exists, then the expression is said to be unsatisfiable. Otherwise, the expression is satisfiable, and the assignment is a solution to the SAT problem. If the Boolean expression is a conjunction (AND) of clauses, then we call it a *Conjunctive Normal Form (CNF)*. Since netlists composed of logic gates can be converted into CNF easily, SAT has been used extensively to solve circuit design and verification problems.

SAT is the first known NP-complete problem [56]. Fortunately, many practical SAT problems can be solved by modern solvers such as MiniSat [61], GRASP [98] and zChaff [103]. However, these solvers still cannot handle many important problems, and more research on this problem is being conducted.

3.2.2 Bounded Model Checking

Bounded Model Checking (BMC) [16] is a formal method which can prove or disprove properties of bounded length in a design, frequently using SAT solving techniques to achieve this goal. A high-level flow of the algorithm is given in Figure 3.2. The central idea of BMC is to "unroll" a given sequential circuit k times to generate a combinational circuit that has behavior equivalent to k clock cycles of the original circuit. In the process of unrolling, the circuit's memory elements are eliminated, and the signals that feed them at cycle i are connected directly to the memory elements' output signals at cycle $i - 1$. In CNF-based SAT, the resulting combinational circuit is converted to a CNF formula C. The property to be proved is also complemented and converted to CNF form ($\overline{p}$). These two formulas are conjoined and the resulting SAT instance I is fed into a SAT solver. If a satisfiable assignment is found for I, then the assignment describes a counterexample that falsifies the (bounded) property, otherwise the property holds true.

```
1  SAT-BMC(circuit, property, maxK)
2  p̄= CNF(!property);
3  for k=1 to maxK do
4     C = CNF(unroll(circuit, k));
5     I = C ∧ p̄;  //SAT instance
6     if (I is satisfiable)
7        return (SAT solution);
```

Figure 3.2. Pseudo-code for bounded model checking.

3.2.3 Symbolic Simulation

The basic idea behind symbolic simulation is similar to that of logic simulation [14]. Unlike logic simulation, however, Boolean variables are simulated instead of constant scalar values. For example, simulating "A AND B" will produce a Boolean expression representing "A AND B" instead of a Boolean value.

In symbolic simulation, a new symbol is injected to each primary input at each cycle. The simulator then produces Boolean expressions at the outputs of the circuit using the injected symbols. Since each symbol implicitly represents both the values 1 and 0, the generated Boolean expressions represent all possible input sequences. As a result, if the design has n inputs, symbolic simulation can produce outputs representing all 2^n input patterns in one single step. Traditionally, *Binary Decision Diagrams (BDDs)* [20] have been used to represent the Boolean expressions due to their flexibility in Boolean manipulations. Recently, symbolic simulators using CNF to represent the Boolean expressions have also been developed [147].

The verification power of symbolic simulation is similar to that of BMC: it can be used to prove properties within a bounded number of cycles or disprove a property; however, it cannot prove a property that considers an indefinite number of cycles. For example, symbolic simulation can falsify a property like "c1 is always equal to 0", or it can prove a property like "c1 always becomes 1 three cycles after a1 is set to 0". Nonetheless, it cannot prove a property that says "c1 is always equal to 0".

3.2.4 Reachability Analysis

Reachability analysis is also called *symbolic traversal* or *least fix-point computation*. It tries to solve the following problem: given a *Finite State Machine (FSM)* description of a sequential digital circuit, find all the reachable states from a set of initial states. Its algorithmic flow is shown in Figure 3.3. In the algorithm, R is a set of reached states, I is the set of initial states, and Δ is the transition function for the FSM (i.e., it maps each (state, input) to a next state). We use subscript t to represent the cycle at which the current computation takes place. The Img function used in the algorithm calculates the forward image of the given states and transition functions. To this end, Coudert et al. [57] provide an efficient algorithm for forward-image computation.

```
1  t= 0;
2  R_t = I; // Start from initial state
3  repeat
4     R_{t+1} = R_t ∪ Img(R_t, Δ); // Compute forward image
5  until (R_{t+1} == R_t); // Repeat until a fix point is reached
```

Figure 3.3. The algorithmic flow of reachability analysis.

Reachability analysis possesses greater verification power than BMC and symbolic simulation in that it can prove properties that consider an infinite number of cycles. To prove properties using reachability analysis, we first identify the set of states P' that do not satisfy the property. Next, we compute the reachable set R. If we found that $R \cap P'$ is empty, then the property holds; otherwise, the property can be violated and there is a bug.

Although reachability analysis is powerful, representing the states is challenging because the number of possible states grows exponentially with the number of state bits. Although BDDs have been shown to be effective in encoding the states, their scalability is still limited. To address this problem, several different approaches have been proposed, including abstraction, parameterization and mixed traversal algorithms [112].

3.2.5 Equivalence Checking

The purpose of equivalence checking is to prove that two circuits exhibit exactly the same behavior. There are two types of equivalency between two circuits: combinational and sequential. Given identical input vectors, combinational equivalency requires both circuits to produce exactly the same responses at their primary outputs and register boundaries. On the other hand, sequential equivalency only requires the responses at primary outputs to be identical.

The basic procedure to perform combinational equivalence checking between two circuits works as follows. First, the inputs/outputs to the registers are broken into primary outputs/inputs of the circuits. Next, a miter is added between each pair of corresponding outputs, where a miter is a circuit consisting of an XOR gate combining the pair of outputs. Third, the corresponding inputs between two circuits are connected to the same signal sources. After inserting these constructs, the equivalence checker then tries to find an input pattern that makes the output of any of the inserted miters 1. If no such pattern can be found, then the two circuits are equivalent; otherwise they are not equivalent and the pattern is a counterexample. The equivalence checker can be implemented using BDDs or CNF-SAT. Techniques that improve this basic procedure have also been proposed, for example [85].

BMC can be used to perform sequential equivalence checking up to a certain number of cycles C, and it works as follows. Given two circuits, they are first unrolled C times. Next, the primary inputs of both circuits for each unrolled copy are connected, and the circuits are constrained using their initial states. Miters are then added to the unrolled primary outputs between both circuits. BDDs or CNF-SAT can then be used to perform the checking. If a sequence of patterns exists that can make the output of any miter 1, then the circuits are not sequentially equivalent, and the sequence becomes a counterexample.

3.3 Design for Debugging and Post-Silicon Metal Fix

Post-silicon debugging is considerably different from pre-silicon debugging because of its special physical constraints. In particular, observing, controlling and changing any circuit component post-silicon is very difficult. To address this problem, existing techniques focus on improving the observability and controllability of the silicon die. In this section we describe the most commonly-used DFD construct, *scan chains*. Next, we introduce the *Focused Ion Beam (FIB)* technique that supports post-silicon metal fix.

3.3.1 Scan Chains

Without special constructs, only the values of a circuit's primary inputs and outputs can be observed and controlled. Therefore, modern chips often use scan chains [23] to improve the design's observability and controllability. The basic idea behind scan chains is to employ sequential elements that have a serial shift capability so that they can be connected to form long shift registers. The scan-chain elements can then operate like primary inputs or outputs during debugging, which can greatly improve the controllability and observability of the circuit's internal signals.

3.3.2 Post-Silicon Metal Fix via Focused Ion Beam

Focused Ion Beam (FIB) is a technique that uses a focused beam of gallium ions [99]. Gallium is chosen because it is easy to liquefy and ionize. After gallium is liquefied, a huge electric field causes ionization and field emission of the gallium atoms, and the ions are focused onto the target by electrostatic lens. When the high-energy gallium ions strike their target, atoms will be sputtered from the surface of the target. Because of this, FIB is often used as a micro-machining tool to modify materials at the nanoscale level. In the semiconductor industry, FIB can be applied to modify an existing silicon die. For example, it can cut a wire or deposit conductive material in order to make a connection. However, it cannot create new transistors on a silicon die.

To remove unwanted materials from a silicon die, *ion milling* is used. When an accelerated ion hits the silicon die, the ion loses its energy by scattering the electrons and the lattice atoms. If the energy is higher than the binding energy of the atoms, the atoms will be sputtered from the surface of the silicon die. To complement material removal, *ion-induced deposition* is used to add new materials to a silicon die. In the process, a precursor gas, often an organometallic, is directed to and absorbed by the surface of the die. When the incident ion beam hits the gas molecule, the molecule dissociates and leaves the metal constituent as a deposit. Similarly, insulator can also be deposited on the die. Since impurities such as gallium ions may be trapped by the deposited materials, the conductivity/resistivity of the deposited metal/insulator tends to be worse than that produced using the regular manufacturing process. Fortunately, this phenomenon does not pose serious challenges in post-silicon debugging because the changes made are typically small.

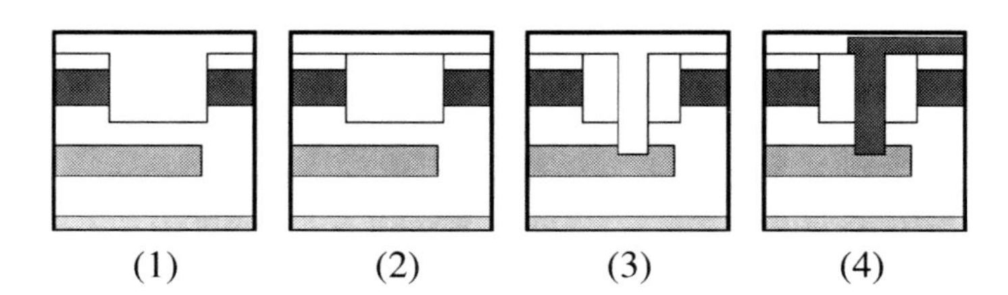

Figure 3.4. Schematic showing the process to connect to a lower-level wire through an upper-level wire: (**a**) a large hole is milled through the upper level; (**b**) the hole is filled with SiO_2; (**c**) a smaller hole is milled to the lower-level wire; and (**d**) the hole is filled with new metal. In the figure, whitespace is filled with SiO_2, and the dark blocks are metal wires.

FIB can either cut or reconnect top-level wires. Changing metallic wires at lower levels, however, is a much more elaborate process. To achieve this, a large hole is first milled through the upper-level wires to expose the lower-level wire, then the hole is filled with oxide for insulation. Next, a new smaller hole is milled through the refilled oxide, and metal is deposited down to the lower level. The affected upper-level wires may need to be reconnected in a similar way. An illustration of the process is shown in Figure 3.4.

PART II

FOGCLEAR METHODOLOGIES AND THEORETICAL ADVANCES IN ERROR REPAIR

Chapter 4

CIRCUIT DESIGN AND VERIFICATION METHODOLOGIES

In this chapter we describe the FogClear methodologies that automate the IC verification and debugging flows, including front-end design, back-end logic design, back-end physical design and post-silicon debugging.

4.1 Front-End Design

Our FogClear front-end methodology automates the functional error correction process, and it works as follows. Given a bug trace and the RTL model that fails verification, *Butramin* (see Chapter 8) is used to minimize the bug trace, and then the minimized bug trace is analyzed by the *REDIR* framework (see Section 9.2) to produce a repaired RTL model. The repaired RTL model is verified again to make sure no new bugs are introduced by the fix. This process keeps repeating until the model passes verification. The FogClear front-end design flow is shown in Figure 4.1, where the "Automatic functional error correction" block replaces the "Manual functional error correction" block in Figure 2.1. By automating the error diagnosis and correction process, engineers' time can be saved, and design quality can be improved.

4.2 Back-End Logic Design

Fixing errors at the gate level is more difficult than at the RTL because engineers are unfamiliar with the synthesized netlists. In order to address this problem, our FogClear design flow automatically repairs the gate-level netlist. As shown in Figure 4.2, it is achieved by analyzing the counterexamples returned by the verification engine using the *CoRé* framework (see Chapter 5 and Section 9.1). This framework automates the gate-level error-correction process and thus saves engineers' time and effort.

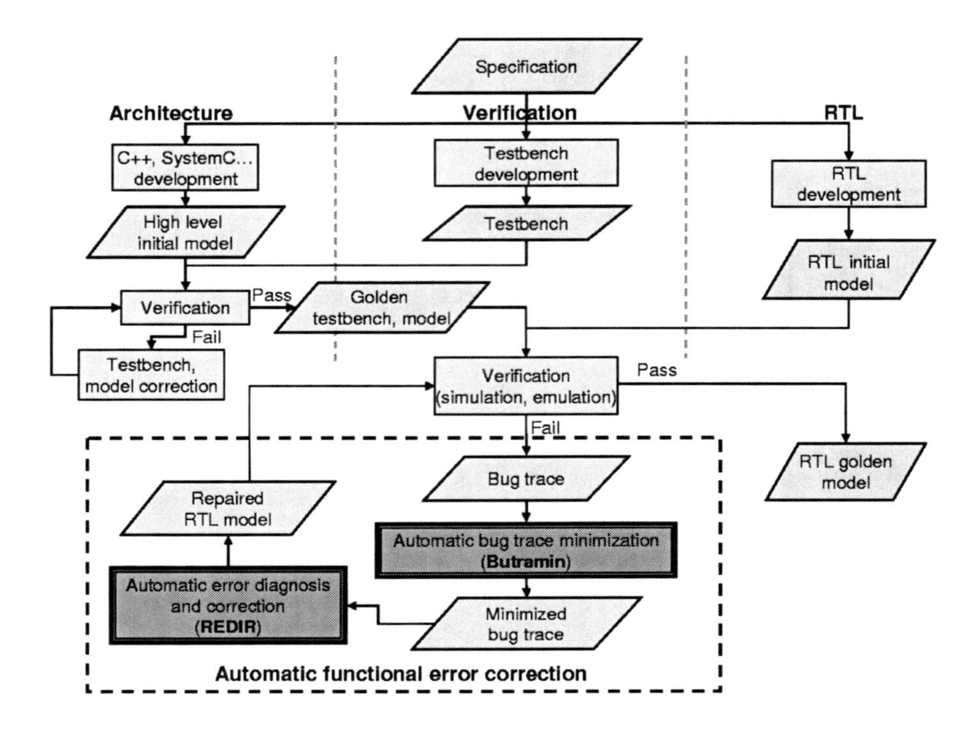

Figure 4.1. The FogClear front-end design flow.

4.3 Back-End Physical Design

Due to the growing dominance of interconnect in delay and power of modern designs, tremendous physical synthesis effort and even more powerful optimizations such as retiming are required. Given that bugs still appear in many EDA tools today [9], it is important to verify the correctness of the performed optimizations. Traditional techniques address this verification problem by checking the equivalence between the original design and the optimized version. This approach, however, only verifies the equivalence of two versions of the design after a number, or possibly all, of the transformations and optimizations have been completed. Unfortunately, such an approach is not sustainable in the long term because it makes the identification, isolation, and correction of errors introduced by the transformations extremely difficult and time-consuming. On the other hand, performing traditional equivalence checking after each circuit transformation is too demanding. Since functional correctness is the most important aspect of high-quality designs, a large amount of effort is currently devoted to verification and debugging, expending resources that could have otherwise been dedicated to improve other aspects of performance. To this end, verification has become the bottleneck

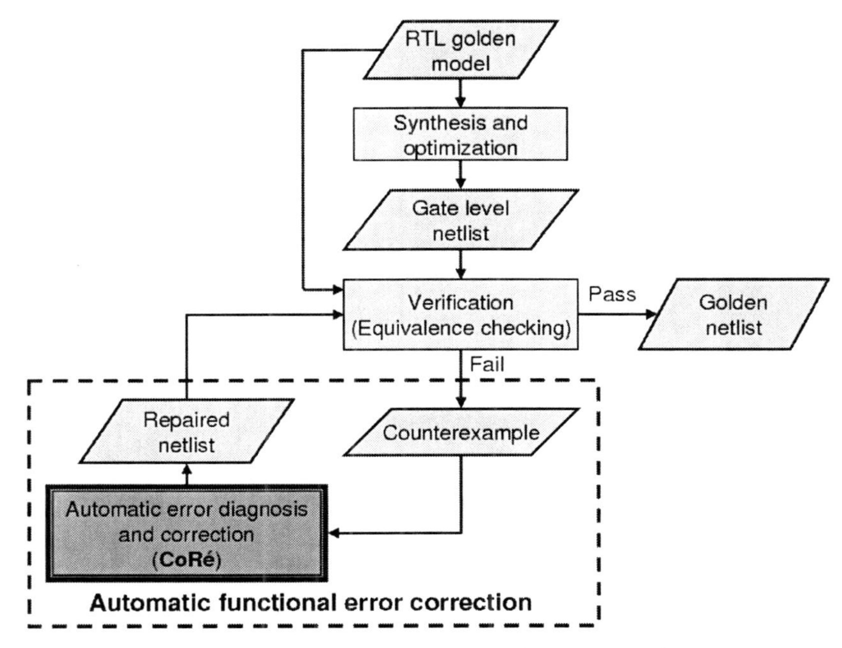

Figure 4.2. The FogClear back-end logic design flow.

that limits achievable optimizations and the features that can be included in a design [49], slowing down the evolution of the overall quality of electronic designs.

The FogClear back-end physical design flow shown in Figure 4.3 addresses this problem using an incremental verification system called *InVerS*, which will be described in detail in Chapter 10. InVerS relies on a metric called *similarity factor* to point out the changes that might have corrupted the circuit. Since this metric is calculated by fast simulation, it can be applied after every circuit modification, allowing engineers to know when a bug might have been introduced and traditional verification should be performed. When the similarity factor indicates a potential problem, traditional verification should be performed to check the correctness of the executed circuit modification. If verification fails, the CoRé framework can be used to repair the errors. Alternatively, the errors can also be fixed by reversing the performed modification.

As Section 10.3 shows, the InVerS system has high accuracy and can catch most errors. However, it is still possible that a few errors may escape incremental verification and be found in the full-fledged post-layout verification. When this happens, the post-silicon error-repair techniques that we describe in the next section can be used to repair the layout and fix the errors.

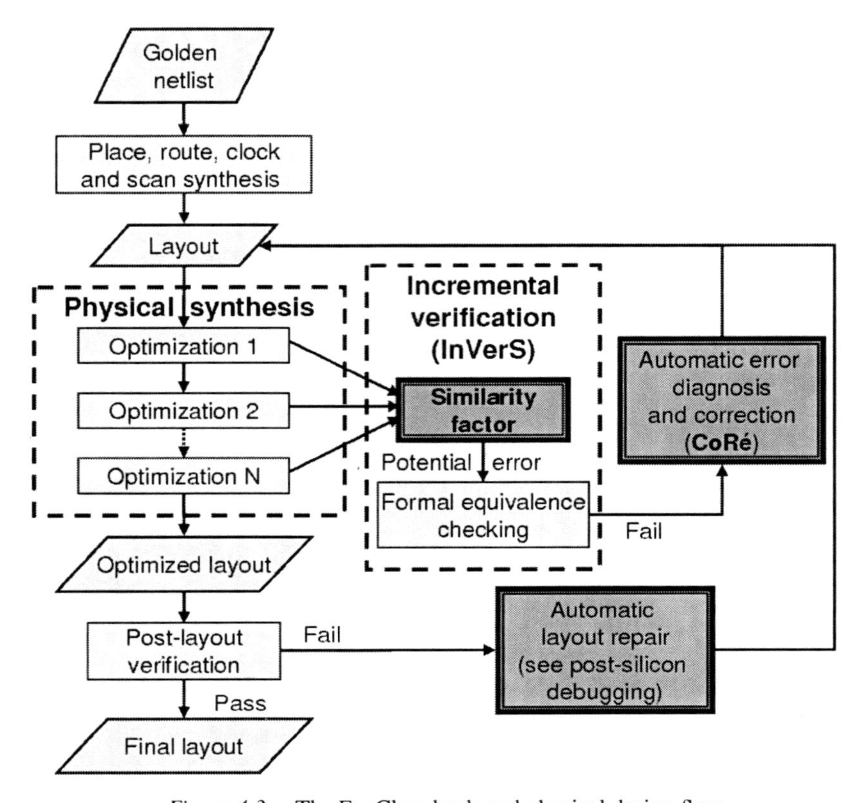

Figure 4.3. The FogClear back-end physical design flow.

4.4 Post-Silicon Debugging

Figure 4.4 shows our FogClear methodology which automates post-silicon debugging. When post-silicon verification fails, a bug trace is produced. Since silicon dies offer simulation speed orders of magnitude faster than that provided by logic simulators, constrained-random testing is used extensively, generating extremely long bug traces. To simplify error diagnosis, we also apply bug trace minimization in our methodology to reduce the complexity of traces using the Butramin technique.

After a bug trace is simplified, we simulate the trace with a logic simulator using the source netlist for the design layout. If simulation exposes the error, then the error is functional, and PAFER is used to generate a repaired layout; otherwise, the error is electrical. Currently, we still require separate error diagnosis steps to find the cause of an electrical error. For example, the techniques proposed by Killpack [81] can be used to diagnose electrical errors. After the cause of the error is identified, we check if the error can be repaired by ECO routing. If so, we apply existing ECO routing tools such as those in [140];

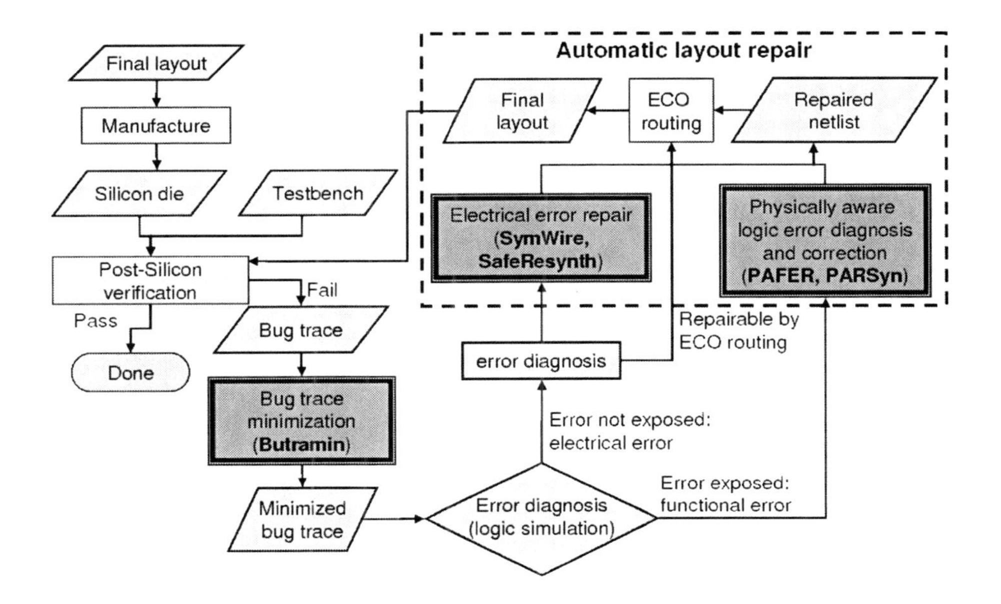

Figure 4.4. The FogClear post-silicon debugging flow.

otherwise, we use SymWire or SafeResynth to change the logic and wire connections around the error spot in order to fix the problem. The layout generated by SymWire or SafeResynth is then routed by an ECO router to produce the final repaired layout. This layout can be used to fix the silicon die for further verification. A more detailed description on the components used in our flow is given in Chapter 11.

Chapter 5

COUNTEREXAMPLE-GUIDED ERROR-REPAIR FRAMEWORK

In this chapter we present a resynthesis framework, called CoRé, that automatically corrects errors in combinational gate-level designs. The framework is based on a new simulation-based abstraction technique and utilizes resynthesis to modify the functionality of a circuit's internal nodes to match the correct behavior. Compared with previous solutions, CoRé is more powerful in that: (1) it can fix a broader range of error types because it is not bounded by specific error models; (2) it derives the correct functionality from simulation vectors, without requiring golden netlists; and (3) it can be applied with a broad range of verification flows, including formal and simulation-based. In this chapter, we first provide required background. Next, we present our CoRé framework that addresses the gate-level error-repair problem.

5.1 Background

In CoRé we assume that an input design, with one or more bugs, is provided as a Boolean network. We strive to correct its erroneous behavior by regenerating the functionality of incorrect nodes. This section starts by defining some terminology and then overviews relevant background.

5.1.1 Bit Signatures

DEFINITION 5.1 *Given a node t in a Boolean network, whose function is f, as well as input vectors x_1, x_2 ... x_k. We define the* signature *of node t, s_t, as $(f(x_1), ..., f(x_k))$, where $f(x_i) \in \{0,1\}$ represents the output of f given an input vector x_i.*

Our goal is to modify the functions of the nodes responsible for the erroneous behavior of a circuit via resynthesis. In this context, we call a node to be resynthesized the *target node*, and we call the nodes that we can use as inputs to the newly synthesized node (function) the candidate nodes. Their corresponding signatures are called the *target signature* and the *candidate signatures*, respectively.

Given a target signature s_t and a collection of candidate signatures s_{c_1}, s_{c_2},...,s_{c_n}, we say that s_t can be resynthesized by s_{c_1}, s_{c_2},...,s_{c_n} if s_t can be expressed as $s_t= f(s_{c_1}, s_{c_2}, ..., s_{c_n})$, where $f(s_{c_1}, s_{c_2}, ..., s_{c_n})$ is a vector Boolean function called the *resynthesis function*. We also call a netlist that implements the resynthesis function the *resynthesized netlist*.

5.1.2 Don't-Cares

When considering a subnetwork within a large Boolean network, *Don't-Cares (DCs)* are exploited by many synthesis techniques because they provide additional freedom for optimizations. *Satisfiability Don't-Cares (SDCs)* occur when certain combinations of input values do not occur for the subnetwork, while *Observability Don't-Cares (ODCs)* occur when the output values of the subnetwork do not affect any primary output. As we show in Section 5.2.1, our CoRé framework is able to utilize both SDCs and ODCs.

5.1.3 SAT-Based Error Diagnosis

The error-diagnosis technique used in our CoRé framework is based on the work by Smith et al. [125]. Given a logic netlist, a set of test vectors and a set of correct output responses, this technique will return a set of wires, also called *error sites*, along with their values for each test vector that can correct the erroneous output responses. Our CoRé framework then corrects design errors by resynthesizing the error sites using the corrected values as the target signatures. In Smith's error-diagnosis technique, three components are added to the netlist, including (1) multiplexers, (2) test-vector constraints, and (3) cardinality constraints. The whole circuit is then converted to CNF, and a SAT solver is used to perform error diagnosis. These components are added temporarily for error diagnosis only and will not appear in the netlist produced by CoRé. They are described in detail below.

A multiplexer is added to every wire to model the correction of the erroneous netlist. When the select line is 0, the original driver of the wire is used. When the select line is 1, the multiplexer chooses a new signal source instead, and the values applied by the new source will correct the erroneous output responses. An example of the multiplexer is given in Figure 5.1(a). A variable, v_i, is introduced for every multiplexer to model the new source to the wire.

Test-vector constraints are used to force the erroneous netlist to produce correct output responses for the test vectors. Obviously, the netlist can produce correct output responses only if a subset of the select lines of the added multiplexers are set to 1, allowing the corresponding new signal sources to

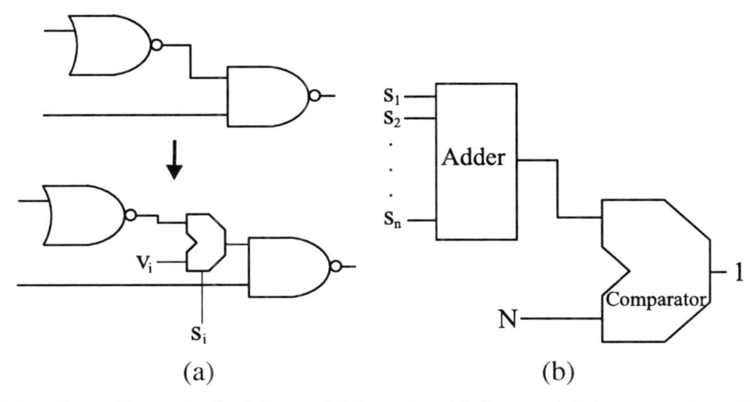

Figure 5.1. Error diagnosis. In (**a**) a multiplexer is added to model the correction of an error, while (**b**) shows the error cardinality constraints that limit the number of asserted select lines to N.

generate logic values that agree with the signal sources which produce the correct responses. These constraints are implemented by duplicating one copy of the multiplexer-enriched netlist for each test vector. The inputs of the copy are controlled by the test vector, and its outputs are constrained by the correct responses.

Cardinality constraints restrict the number of select lines that can be set to 1 simultaneously. This number also represents the number of error sites in the netlist. The cardinality constraints are implemented by an adder which performs a bitwise addition of the select lines, and a comparator which forces the sum of the adder to be N, as shown in Figure 5.1(b). Initially, N is set to 1, and error diagnosis is performed by incrementing N until a solution is found.

5.1.4 Error Model

To reduce the complexity of error diagnosis and correction, several error models have been introduced. These models classify common design errors in order to reduce the difficulty of repairing them. Here we describe a frequently used model formulated by D. Nayak [104], which is based on Abadir's model [1].

In the model, type "a" (wrong gate) mistakenly replaces one gate type by another one with the same number of inputs; types "b" and "c" (extra/missing wire) use a gate with more or fewer inputs than required; type "d" (wrong input) connects a gate input to a wrong signal; and types "e" and "f" (extra/missing gate) incorrectly add or remove a gate. An illustration of the model is given in Figure 5.2.

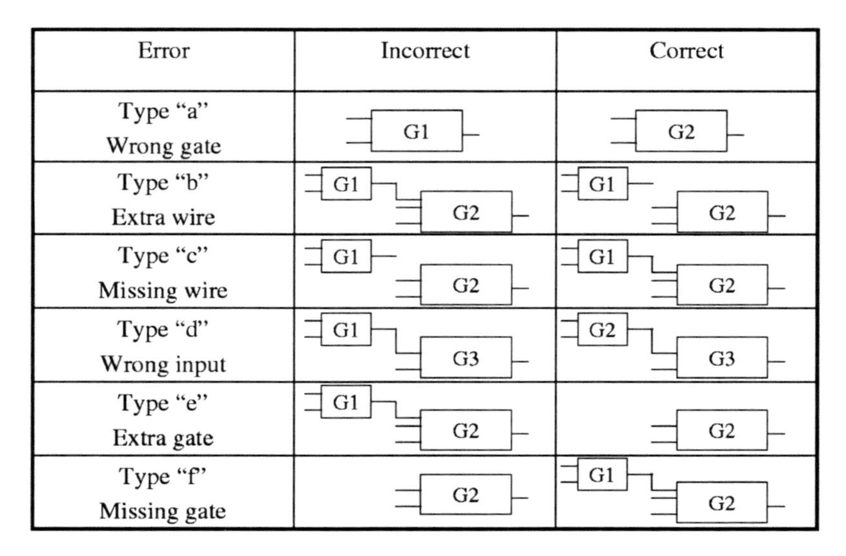

Figure 5.2. Errors modeled by Abadir et al. [1].

5.2 Error-Correction Framework for Combinational Circuits

For the discussion in this section we restrict our analysis to combinational designs. In this context, the correctness of a circuit is simply determined by the output responses under all possible input vectors. We will show in Section 9.1 how to extend the framework to sequential designs.

CoRé, our error-correction framework, relies on simulation to generate signatures, which constitute our abstract model of the design and are the starting point for the error diagnosis and resynthesis algorithms. After the netlist is repaired, it is checked by a verification engine. If verification fails, possibly due to new errors introduced by the correction process, new counterexamples are generated and used to further refine the abstraction. Although in our implementation we adopted Smith's error-diagnosis technique [125] due to its scalability, alternative diagnosis techniques can be used as well.

5.2.1 The CoRé Framework

In CoRé, an input test vector is called a *functionality-preserving vector* if its output responses comply with the specification, and the vector is called an *error-sensitizing vector* if its output responses differ. *Error-sensitizing vectors* are often called *counterexamples*.

The algorithmic flow of CoRé is outlined in Figure 5.3. The inputs to the framework are the original buggy netlist (CKT_{err}), the initial functionality-preserving vectors ($vectors_p$) and the initial error-sensitizing vectors

($vectors_e$). The output is the rectified netlist CKT_{new}. The framework first performs error diagnosis to identify error locations and the correct values that should be generated for those locations so that the error-sensitizing vectors could produce the correct output responses. Those error locations constitute the *target nodes* for resynthesis. The bits in the target nodes' signatures that correspond to the error-sensitizing vectors must be corrected according to the diagnosis results, while the bits that correspond to the functionality-preserving vectors must remain unchanged. If we could somehow create new combinational netlist blocks that generate the required signatures at the target nodes using other nodes in the Boolean network, we would be able to correct the circuit's errors, at least those that have been exposed by the error-sensitizing vectors. Let us assume for now that we can create such netlists (techniques to this end will be discussed in the next chapter), producing the new circuit CKT_{new} (line 4). CKT_{new} is checked at line 5 using the verification engine. When verification fails, new error-sensitizing vectors for CKT_{new} will be returned in $counterexample$. If no such vector exists, the circuit has been successfully corrected and CKT_{new} is returned. Otherwise, CKT_{new} is abandoned, while *counterexample* is classified either as error-sensitizing or functionality-preserving with respect to the original design (CKT_{err}). If $counterexample$ is error-sensitizing, it will be added to $vectors_e$ and be used to rediagnose the design. CKT_{err}'s signatures are then updated using $counterexample$. By accumulating both functionality-preserving and error-sensitizing vectors, CoRé will avoid reproposing the same wrong correction; hence guaranteeing that the algorithm will eventually complete. Figure 5.4 illustrates a possible execution scenario with the flow that we just described.

```
CoRé(CKT_err, vectors_p, vectors_e, CKT_new)
 1  compute_signatures(CKT_err, vectors_p, vectors_e);
 2  fixes= diagnose(CKT_err, vectors_e);
 3  foreach fix ∈ fixes
 4    CKT_new= resynthesize(CKT_err, fix);
 5    counterexample=verify(CKT_new);
 6    if (counterexample is empty)  return CKT_new;
 7    else if (counterexample is error-sensitizing for CKT_err)
 8        vectors_e = vectors_e ∪ counterexample;
 9        fixes= rediagnose(CKT_err, vectors_e);
10    update_signatures(CKT_err, counterexample);
```

Figure 5.3. The algorithmic flow of CoRé.

SDCs are exploited in CoRé by construction because simulation can only produce legal signatures. To utilize ODCs, we simulate the complement signature of the target node and mark the bit positions whose changes do not propagate to any primary output as ODCs: those positions are not considered

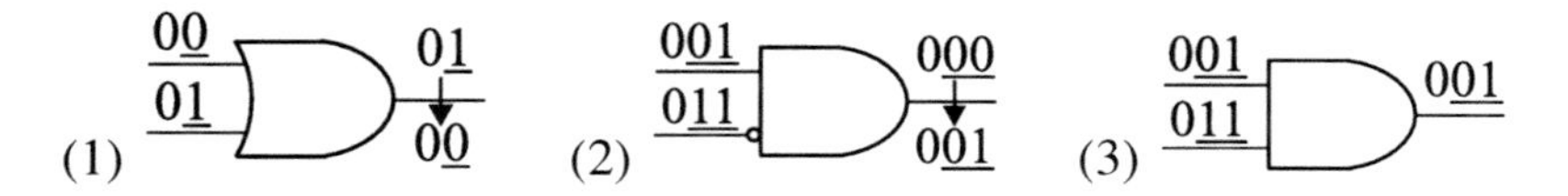

Figure 5.4. Execution example of CoRé. Signatures are shown above the wires, where underlined bits correspond to *error-sensitizing vectors*. (**1**) The gate was meant to be AND but is erroneously an OR. Error diagnosis finds that the output of the 2nd pattern should be 0 instead of 1; (**2**) the first resynthesized netlist fixes the 2nd pattern, but fails further verification (the output of the 3rd pattern should be 1); (**3**) the counterexample from step 2 refines the signatures, and a resynthesized netlist that fixes all the test patterns is found.

during resynthesis. Note that if a diagnosis contains multiple error sites, the sites that are closer to primary outputs should be resynthesized first so that the downstream logic of a node is always known when ODCs are calculated.

5.2.2 Analysis of CoRé

To achieve the required scalability to support the global implications of error correction, CoRé uses an abstraction-refinement scheme: signatures provide an abstraction of the Boolean network for resynthesis because they are the nodes' partial truth tables (all unseen input vectors are considered as DCs), and the abstraction is refined by means of the counterexamples that fail verification. The following proposition shows that in theory, CoRé can eventually always produce a netlist which passes verification. However, as it is the case for most techniques based on abstraction and refinement, the framework may time-out before a valid correction is found in practice. The use of high-quality test vectors [130] is effective in alleviating this potential problem.

PROPOSITION 1 *Given a buggy combinational design and a specification that defines the output responses of each input vector, the CoRé algorithm can always generate a netlist that produces the correct output responses.*

PROOF 1 *Given a set of required "fixes", the resynthesis function of CoRé can always generate a correct set of signatures, which in turn produce correct responses at primary outputs. Observe that each signature represents a fragment of a signal's truth table. Therefore, when all possible input patterns are applied to our CoRé framework, the signatures essentially become complete truth tables, and hence define all the terms required to generate correct output responses for any possible input stimulus. In CoRé, all the counterexamples that fail verification are used to expand and enhance the set of signatures. Each correction step of CoRé guarantees that the output responses of the input patterns seen so far are correct, thus any counterexample must be new. However, since the number of distinct input patterns is finite (at most 2^n for an n-input circuit), eventually no new vector can be generated, guaranteeing* *that the algorithm will complete in a finite number of iterations. In practice, we find that a correct design can often be found in a few iterations.*

5.2.3 Discussions

Several existing techniques, such as those in [130], also use simulation to identify potential error-correction options and rely on further simulation to prune unpromising candidates. Compared with these techniques, the framework described in this section is more flexible because it performs abstraction and refinement on the design itself. As a result, this framework can easily adopt new error diagnosis or correction techniques. For example, our error-correction engine can be easily replaced by any synthesis tool that can handle truth tables or cubes. Most existing techniques, however, do not have this flexibility. On the other hand, recent work by Safarpour et al. [116] provides another abstraction-refinement error-repair methodology for sequential circuits by replacing a fraction of the registers with primary inputs. Their methodology can be used to accelerate the diagnosis process in our error-repair method for sequential circuits, which is described in Section 9.1.

5.2.4 Applications

CoRé can be used whenever the output responses of a netlist need to be changed. We now develop applications of our techniques in three different verification contexts.

Application 1: combinational equivalence checking and enforcement. This application fixes an erroneous netlist so that it becomes equivalent to a golden netlist. In this application, the verification engine is an equivalence checker. Test vectors on which the erroneous circuit and the golden model agree are functionality-preserving vectors, and the remaining test vectors are error-sensitizing. Initial vectors can be obtained by random simulation or equivalence checking.

Application 2: fixing errors found by simulation. This application corrects design errors that break a regression test. In this application, the verification engine is the simulator and the regression suite. Test vectors that break the regression are error-sensitizing vectors, and all other vectors are functionality-preserving vectors. Initial vectors can be obtained by collecting the inputs applied to the netlist while running the regression.

Application 3: fixing errors found by formal verification. This application assumes that a formal tool proves that a property can be violated, and the goal is to fix the netlist to prevent the property from being violated. In this application, counterexamples returned by the tool are error-sensitizing vectors.

Chapter 6

SIGNATURE-BASED RESYNTHESIS TECHNIQUES

The basis for CoRé's resynthesis solution is the signature available at each internal circuit node, where the signature of a node is essentially its partial truth table. The resynthesis problem is formulated as follows: given a target signature, find a resynthesized netlist that generates the target signature using the signatures of other nodes in the Boolean network as inputs. In this Chapter, we first describe the concept of *Pairs of Bits to be Distinguished (PBDs)*, which compactly encode resynthesis information. Next, we describe our *Distinguishing-Power Search (DPS)* and *Goal-Directed Search (GDS)* resynthesis techniques that are based on signatures.

6.1 Pairs of Bits to be Distinguished (PBDs)

In this section we propose the concepts of *Pairs of Bits to be Distinguished (PBDs)* and *distinguishing power*. PBDs can be derived easily using signatures and compactly encode the information required for resynthesis. A similar concept, *Sets of Pairs of Functions to be Distinguished (SPFDs)* [123, 141, 142], is also described.

6.1.1 PBDs and Distinguishing Power

Recall that a signature s is a collection of the corresponding node's simulation values. In this chapter, we use $s[i]$ to denote the i-th bit of signature s. The goal of error correction is to modify the functions of the nodes responsible for the erroneous behavior of a circuit via resynthesis. In this context, we call a node to be resynthesized the *target node*, and we call the nodes that we can use as inputs to the newly synthesized node (function) the *candidate nodes*. Their corresponding signatures are called the *target signature* and the *candidate signatures*, respectively.

The proposition below states that a sufficient and necessary condition for a resynthesis function to exist is that, whenever two bits in the target signature are distinct, then such bits need to be distinct in at least one of the candidate signatures. This proposition is a special case of Theorem 5.1 in [101], where the minterms appearing in signatures represent the care-terms and all other minterms are *Don't-Cares (DCs)*.

PROPOSITION 2 *Consider a collection of candidate signatures, s_{c_1}, s_{c_2},...,s_{c_n}, and a target signature, s_t. Then a resynthesis function f, where $s_t = f(s_{c_1}, s_{c_2},...,s_{c_n})$, exists if and only if no bit pair $\{i,j\}$ exists such that $s_t[i] \neq s_t[j]$ but $s_{c_k}[i] = s_{c_k}[j]$ for all $1 \leq k \leq n$.*

In this work we call a pair of bits $\{i,j\}$ in s_t, where $s_t[i] \neq s_t[j]$, a *Pair of Bits to be Distinguished (PBD)*. Based on Prop. 2, we say that the PBD $\{i,j\}$ can be *distinguished* by signature s_{c_k} if $s_{c_k}[i] \neq s_{c_k}[j]$. We define the *Required Distinguishing Power (RDP)* of the target signature s_t, $RDP(s_t)$, as the set of PBDs that need to be distinguished. We also define the *Distinguishing Power (DP)* of a candidate signature s_{c_k} with respect to the target signature s_t, $DP(s_{c_k}, s_t)$, as the set of PBDs in s_t that can be distinguished by s_{c_k}. With this definition, Prop. 2 can be restated as "a resynthesis function, f, exists if and only if $RDP(s_t) \subseteq \cup_{k=1}^{n} DP(s_{c_k}, s_t)$".

6.1.2 Related Work

SPFD [123, 141, 142] is a relatively new data structure that encodes resynthesis information and allows the use of DCs. An SPFD [141] R_t for a target node t, represented as $\{(g_{1a}, g_{1b}), (g_{2a}, g_{2b}), ..., (g_{na}, g_{nb})\}$, denotes a set of pairs of functions that must be distinguished. In other words, for each pair $(g_{ia}, g_{ib}) \in R_t$, the minterms in g_{ia} must produce a different value from the minterms in g_{ib} at the output of t. Assume that node t has m fanins, $c_1...c_m$, and their SPFDs are $R_{c_1}...R_{c_m}$, then according to [142]:

$$R_t \subseteq \cup_{i=1}^{m} R_{c_i} \tag{6.1}$$

In other words, the SPFD of node t is a subset of the union of all the SPFDs of its fanins $c_1...c_m$. Since a function f satisfies an SPFD R_t if and only if for each $(g_{ia}, g_{ib}) \in R_t$, $f(g_{ia}) \neq f(g_{ib})$ [141]. This criterion, combined with Equation 6.1, essentially states that a resynthesis function f exists if and only if all the minterms that need to be distinguished in R_t are distinguished by at least one of its fanins, which is consistent with Prop. 2. As a result, our use of PBDs is equivalent to SPFDs. However, our approach has the following advantages over SPFD-based techniques:

1 PBDs provide a much more compact logic representation than SPFDs. Traditionally, SPFDs are calculated using BDDs and suffer memory explosion problems. Recent work represents SPFDs as graphs [123] and SAT/simulation can be used to calculate SPFDs [101]. This approach is more memory efficient but may become computationally expensive. On the other hand, our approach only uses signatures of nodes. Since each minterm needs only one bit in a signature, our representation is very compact.

2 Calculating PBDs is significantly easier than calculating SPFDs: signatures are generated by simulation, and DCs are calculated by simulating the complement of the target signature. PBDs can then be derived easily by considering only the care-terms in the target signature.

6.2 Resynthesis Using Distinguishing-Power Search

In this section, we first define the *absolute distinguishing power* $|DP(s)|$ of a signature s, and then we propose a *Distinguishing-Power Search (DPS)* technique that uses |DP| to select candidate signatures and generates the required resynthesized netlist.

6.2.1 Absolute Distinguishing Power of a Signature

Absolute distinguishing power provides search guiding and pruning criteria for our resynthesis techniques. To simplify bookkeeping, we reorder bits in every signature so that in the target signature all the bits with value 0 precede the ones with value 1, as in "00...0011...11".

DEFINITION 6.1 *Assume a target signature s_t is composed of x 0s followed by y 1s, we define the* absolute required distinguishing power *of s_t, denoted by $|RDP(s_t)|$, as the number of PBDs in s_t and equals to xy. Moreover, if a candidate signature s_c has p 0s and q 1s in its first x bit positions, and r 0s and s 1s in the remaining y positions, then we define the* absolute distinguish power *of s_c with respect to s_t, denoted by $|DP(s_c, s_t)|$, as the number of PBDs in s_t that can be distinguished by s_c and equals to $ps + qr$.*

The following corollary states a necessary but not sufficient condition to determine whether the target signature can be generated from a collection of candidate signatures.

COROLLARY 6.2 *Consider a target signature s_t and a collection of candidate signatures $s_{c_1}...s_{c_n}$. If s_t can be generated by $s_{c_1}...s_{c_n}$, then $|RDP(s_t)| \leq \sum_{i=1}^{n} |DP(s_{c_i}, s_t)|$.*

6.2.2 Distinguishing-Power Search

Distinguishing-Power Search (DPS) is based on Prop. 2, which states that a resynthesis function can be generated when a collection of candidate signatures covers all the PBDs in the target signature. However, the number of collections satisfying this criterion may be exponential in the number of total signatures. To identify possible candidate signatures effectively, we first select signatures that cover the least-covered PBDs, second those that have high absolute distinguishing power (i.e., signatures that cover the most number of PBDs), and third those that cover any remaining uncovered PBD. For efficiency, we limit the search pool to the 200 nodes which are topologically closest to the target node; however, we may go past this limit when those are not sufficient to cover all the PBDs in the target signature. Finally, we exclude from the pool those nodes that are in the fanout cone of the target node, so that we avoid creating a combinational loop inadvertently.

After the candidate signatures are selected, a truth table for the resynthesis function is built from the signatures, and it is constructed as follows. Note that although we may select more signatures than needed for resynthesis, the logic optimizer we use in the next step is usually able to identify the redundant signatures and use only those which are essential.

1 Each signature is an input to the truth table. The i-th input produces the i-th column in the table, and the j-th bit in the signature determines the value of the j-th row.

2 If the j-th bit of the target signature is 1, then the j-th row is a minterm; otherwise it is a maxterm.

3 All other terms are don't-cares.

Figure 6.1 shows an example of the constructed truth table. The truth table can be synthesized and optimized using existing software, such as Espresso [114] or MVSIS [64]. Note that our resynthesis technique does not require that the support of the target function is known *a priori*, since the correct support will be automatically selected when DPS searches for a set of candidate signatures that distinguish all the PBDs. This is in contrast with other previous solutions which require that the support of the target node to be known before attempting to synthesize the function.

6.3 Resynthesis Using Goal-Directed Search

Goal-Directed Search (GDS) performs an exhaustive search for resynthesized netlists. To reduce the search space, we propose two pruning techniques: the absolute-distinguishing-power test and the compatibility test. Currently, BUFFERs, INVERTERs, and 2-input AND, OR and XOR gates are supported.

The absolute-distinguishing-power test relies on Corollary 6.2 to reject resynthesis opportunities when the selected candidate signatures do not have sufficient absolute distinguishing power. In other words, a collection of candidate

Signature	Truth table				
s_t=0101	s_1	s_2	s_3	s_4	s_t
s_1=1010	1	0	1	0	0
s_2=0101	0	1	1	0	1
s_3=1110	1	0	1	0	0
s_4=0001	0	1	0	1	1
Minimized	0	-	-	-	1

Figure 6.1. The truth table on the right is constructed from the signatures on the left. Signature s_t is the target signature, while signatures s_1 to s_4 are candidate signatures. The minimized truth table suggests that s_t can be resynthesized by an INVERTER with its input set to s_1.

signatures whose total absolute distinguishing power is less than the absolute required distinguishing power of the target signature is not considered for resynthesis.

The compatibility test is based on the controlling values of logic gates. To utilize this feature, we propose three rules, called *compatibility constraints*, to prune the selection of inputs according to the output constraint and the gate being tried. Each constraint is accompanied with a signature. In particular, an *identity constraint* requires the input signature to be identical to the constraint's signature; and a *need-one constraint* requires that specific bits in the input signatures must be 1 whenever the corresponding bits in the constraint's signature are 1. *Identity constraints* are used to encode the constraints imposed by BUFFERs and INVERTERs, while *need-one constraints* are used by AND gates. Similarly, *need-zero constraints* are used by OR gates. For example, if the target signature is 0011, and the gate being tried is AND, then the *need-one constraint* will be used. This constraint will reject signature 0000 as the gate's input because its last two bits are not 1, but it will accept 0111 because its last two bits are 1. These constraints, which propagate from the outputs of gates to their inputs during resynthesis, need to be recalculated for each gate being tried. For example, an *identity constraint* will become a *need-one constraint* when it propagates through an AND gate, and it will become a *need-zero constraint* when it propagates through an OR gate. The rules for calculating the constraints are shown in Figure 6.2.

The GDS algorithm is given in Figure 6.3. In the algorithm, $level$ is the level of logic being explored, $constr$ is the constraint, and C returns a set of candidate resynthesized netlists. Initially, $level$ is set to 1, and $constr$ is *identity constraint* with signature equal to the target signature s_t. Function $update\_constr$ is used to update constraints.

GDS can be used to find a resynthesized netlist with minimal logic depth. This is achieved by calling GDS iteratively, with an increasing value of the $level$ parameter, until a resynthesized netlist is found. However, the pruning constraints weaken with each additional level of logic in GDS. Therefore, the

	Identity	Need-one	Need-zero
INVERTER	S.C.	S.C.+Need-zero	S.C.+Need-one
BUFFER	Constraint unchanged		
AND	Need-one	Need-one	None
OR	Need-zero	None	Need-zero

Figure 6.2. Given a constraint imposed on a gate's output and the gate type, this table calculates the constraint of the gate's inputs. The output constraints are given in the first row, the gate types are given in the first column, and their intersection is the input constraint. "S.C." means "signature complemented."

Function $GDS(level, constr, C)$
1 if $(level == max\_level)$
2 C= candidate nodes whose signatures comply with $constr$;
3 return;
4 foreach $gate \in library$
5 $constr_n$= $update\_constr(gate, constr)$;
6 $GDS(level + 1, constr_n, C_n)$;
7 foreach $c_1, c_2 \in C_n$
8 if $(level > 1$ or $|DP(c_1, s_t)|+|DP(c_2, s_t)|\geq|RDP(s_t)|)$
9 $s_n = calculate\_signature(gate, c_1, c_2)$;
10 if (s_n complies with $constr$)
11 $C = C \cup gate(c_1, c_2)$;

Figure 6.3. The GDS algorithm.

maximum logic depth for GDS is typically small, and we rely on DPS to find more complex resynthesis functions.

Chapter 7

SYMMETRY-BASED REWIRING

Rewiring is a post-placement optimization that reconnects wires in a given netlist without changing its logic function. To this end, symmetry-based rewiring is especially suitable for post-silicon error repair because no transistors will be affected. In light of this, we propose a rewiring algorithm based on functional symmetries in this chapter. In the algorithm, we extract small subcircuits consisting of several gates from the design and reconnect pins according to the symmetries of the subcircuits. We observe that the power of rewiring is determined by the underlying symmetry detector. For example, the rewiring opportunity in Figure 7.1(a) cannot be discovered unless both input and output symmetries can be detected. In addition, a rewiring opportunity such as the one shown in Figure 7.1(b) can only be found if *phase-shift* symmetries [84] can be detected, where a phase-shift symmetry is a symmetry involving negation of inputs and/or outputs. To enhance the power of symmetry detection, we also propose a graph-based symmetry detector that can identify permutational and phase-shift symmetries on multiple input and output wires, as well as their combinations, creating abundant opportunities for rewiring. In this chapter, we apply our techniques for wirelength optimization and observe that it provides wirelength reduction comparable to that achieved by detailed placement. In Chapter 11, we describe how this technique can be applied to repair post-silicon electrical errors.

The remainder of the chapter is organized as follows. Section 7.1 introduces basic principles of symmetry and describes relevant previous work on symmetry detection and circuit rewiring. In Section 7.2 we describe our symmetry-detection algorithm. Section 7.3 discusses the post-placement rewiring algorithm. Finally, we provide experimental results in Section 7.4 and summarize in Section 7.5.

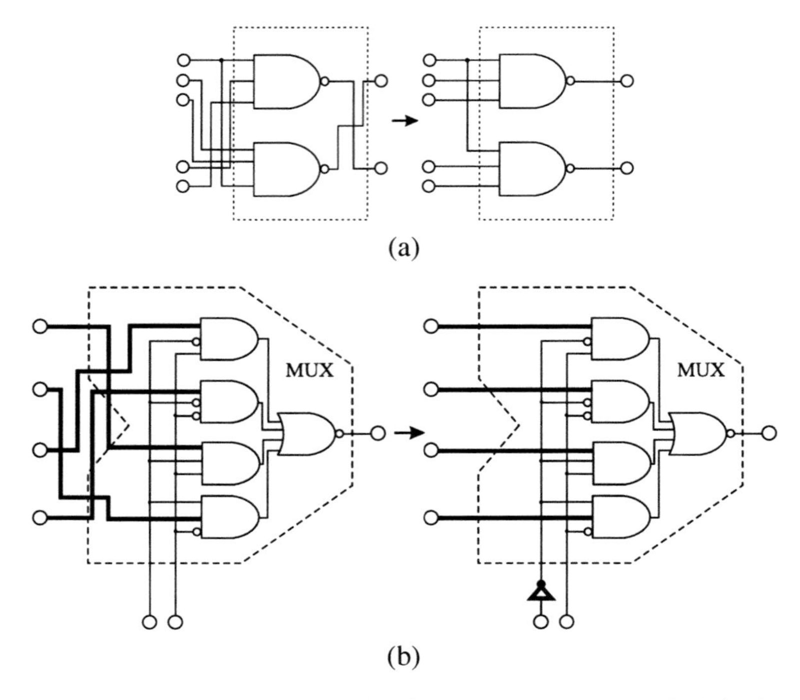

Figure 7.1. Rewiring examples: (**a**) multiple inputs and outputs are rewired simultaneously using pin-permutation symmetry, (**b**) inputs to a multiplexer are rewired by inverting one of the select signals. Bold lines represent changes made in the circuit.

7.1 Background

The rewiring technique described in this chapter is based on symmetry detection. Therefore, in this section, we present background ideas and related work about symmetry detection. Previous work on post-placement rewiring is also discussed.

7.1.1 Symmetries in Boolean Functions

One can distinguish semantic (functional) symmetries of Boolean functions from the symmetries of specific representations (syntactic symmetries). All syntactic symmetries are also semantic, but not vice versa. For example, in function "$o = (x + y) + z$", $x \leftrightarrow z$ is a semantic symmetry because the function will not be changed after the permutation of variables; however, it is not a syntactic symmetry because the structure of the function will be changed. On the other hand, $x \leftrightarrow y$ is both a semantic and syntactic symmetry. In this work we exploit functional symmetries, whose definition is provided below.

DEFINITION 7.1 *Consider a multi-output Boolean function $F : \mathcal{B}^n \rightarrow \mathcal{B}^m$, where*

$$F(i_1...i_n) = < f_1(i_1...i_n), f_2(i_1...i_n)...f_m(i_1...i_n) > . \quad (7.1)$$

*A **functional symmetry** is a one-to-one mapping $s : \mathcal{B}^{(n+m)} \rightarrow \mathcal{B}^{(n+m)}$ such that:*

$$< f_1(i_1...i_n), f_2(i_1...i_n)...f_m(i_1...i_n) >= \quad (7.2)$$
$$< s(f_1)(s(i_1)...s(i_n)), s(f_2)(s(i_1)...s(i_n))...s(f_m)(s(i_1)...s(i_n)) > .$$

In other words, a functional (semantic) symmetry is a transformation of inputs and outputs which does not change the functional relation between them.

EXAMPLE 7.2 *Consider the multi-output function $z = x_1$ XOR y_1 and $w = x_2$ XOR y_2. The variable-permutation symmetries include: (1) $x_1 \leftrightarrow y_1$, (2) $x_2 \leftrightarrow y_2$, (3) $x_1 \leftrightarrow x_2$, $y_1 \leftrightarrow y_2$, and $z \leftrightarrow w$ (all swaps are performed simultaneously). In fact, all the symmetries of this function can be generated from combinations of the symmetries listed above. A set of symmetries with this property are called* symmetry generators. *For example, the symmetry "$x_1 \leftrightarrow y_2$, $y_1 \leftrightarrow x_2$, and $z \leftrightarrow w$" can be generated by applying the symmetries (1), (2) and (3) consecutively.*

While most previous work on symmetry detection focuses on permutations of two variables, Pomeranz [109] and Kravets [84] consider swaps of groups of ordered variables. These swaps are called *higher-order symmetries* in [84]. For example, if variables a, b, c and d in the support of function f satisfy the condition:

$$F(.., a, .., b, .., c, .., d, ..) = F(.., c, .., d, .., a, .., b, ..)$$

then we say that f has a *second-order symmetry* between ordered variable groups (a, b) and (c, d). Such higher-order symmetries are common in realistic designs. For example, in a 4-bit adder, all bits of the two input numbers can be swapped as groups (preserving the order of the bits), but no two input bits in different bit positions are symmetric by themselves. Kravets also introduced phase-shift symmetry as a function-preserving transformation involving the inversion of one or more inputs that do not permute any of the inputs. Our work generalizes this concept by including output symmetries involving inversion in the class of phase-shift symmetries. We also define *composite phase-shift symmetry* as a symmetry which consists of phase-shift and permutational symmetries. In this chapter we commonly refer to composite phase-shift symmetries as just phase-shift symmetries, except for pure phase-shift symmetries which do not include permutations.

EXAMPLE 7.3 *Consider again the multi-output function $z = x_1$ XOR y_1 and $w = x_2$ XOR y_2 given in the previous example. Aside from the pin-swap symmetries discussed in the example, the following phase-shift symmetries also* *exist in the circuit: (1)* $x_2 \leftrightarrow y_2'$*, (2)* $x_1 \leftrightarrow y_1'$*, (3)* $x_2 \leftrightarrow x_2'$ *and* $w \leftrightarrow w'$*, (4)* $x_1 \leftrightarrow x_1'$ *and* $z \leftrightarrow z'$*. Among these symmetries, (1) and (2) are composite phase-shift symmetries because they involve both inversion and permutation of inputs, while (3) and (4) are pure phase-shift symmetries because only inversions of inputs and outputs are used. Note that due to Boolean consistency, a symmetry composed of complement of variables in another symmetry is the same symmetry. For example,* $y_2 \leftrightarrow x_2'$ *is the same as* $x_2 \leftrightarrow y_2'$*.*

7.1.2 Semantic and Syntactic Symmetry Detection

Symmetry detection in Boolean functions has several applications, including technology mapping, logic synthesis, BDD minimization [106] and circuit rewiring [32]. Methods for symmetry detection can be classified into four categories: BDD-based, graph-based, circuit-based and Boolean-matching-based. However, it is relatively difficult to find all symmetries of a Boolean function regardless of the representation used.

BDDs are particularly convenient for semantic symmetry detection because they support abstract functional operations. One naive way to find two-variable symmetries is to compute the cofactors for every pair of variables, say they are v_1 and v_2, and check if $F_{\overline{v_1}v_2} = F_{v_1\overline{v_2}}$ or $F_{\overline{v_1}\,\overline{v_2}}=F_{v_1v_2}$. Recent research [100] indicates that symmetries can be found or disproved without computing all the cofactors and thus significantly speeds up symmetry detection. However, work on BDD-based symmetry detection has been limited to input permutations only.

In this book, symmetry-detection methods that rely on efficient algorithms for the graph-automorphism problem (i.e., finding all symmetries of a given graph) are classified as graph-based. They construct a graph whose symmetries faithfully capture the symmetries of the original object, find its automorphisms (symmetries), and map them back to the original object. Aloul et al. [7] proposed a way to find symmetries for SAT clauses using this approach. The symmetry-detection approach proposed in this book is inspired by their work.

Circuit-based symmetry-detection methods often convert a circuit representing the function in question to a more regular form, where symmetry detection is more practical and efficient. For example, Wang et al. [135] transform the circuit to NOR gates. C.-W. Chang et al. [32] use a more elaborate approach by converting the circuit to XOR, AND, OR, INVERTER and BUFFER first, and then partition the circuit so that each subcircuit is fanout free. Next, they form *supergates* from the gates and detect symmetries for those supergates. Wallace et al. [134] use concepts from Boolean decomposition [15] and convert the circuit to *quasi-canonical forms*, and then input symmetries are recognized from these forms. This technique is capable of finding higher-order symmetries. Another type of circuit-based symmetry detector relies on ATPG and simulation, such as the work by Pomeranz et al. [109]. Although their technique was developed to find both first and higher-order symmetries, they reported experimental results for first-order symmetries only. Therefore, its capability to detect higher-order symmetries is unclear.

Boolean matching is a problem related to symmetry detection. Its purpose is to compute a canonical representation for Boolean functions that are equivalent under negation and permutation of inputs and outputs. Symmetries are implicitly processed by Boolean matching in that all functions symmetric to each other will have the same canonical representation. However, enumerating symmetries from Boolean matching is not straight forward and requires extra work. This topic has been studied by Wu et al. [139] and Chai et al. [28].

Table 7.1. A comparison of different symmetry-detection methods.

Data structure used	Target	Symmetries detected	Main applications	Time complexity
BDD [100]	Boolean functions	All 1st order input symmetries	Synthesis	$O(n^3)$
Circuit – Supergate [32]	Gate-level circuits	1st order input symmetries in supergates, opportunistically	Rewiring, technology mapping	$O(m)$
Circuit – Boolean decomposition [134]	Gate-level circuits	Input and output permutational symmetries, higher-order	Rewiring, physical design	$\Omega(m)$
Circuit – simulation, ATPG [109]	Gate-level circuits	Input, output and phase-shift symmetries, higher-order	Error diagnosis, technology mapping	$\Omega(2^n)$
Boolean matching [28]	Boolean functions	Input, output and phase-shift symmetries, higher-order	Technology mapping	$\Omega(2^n)$
Graph automorphism (our work)	Both (with small number of inputs)	All input, output, phase-shift symmetries and all orders, exhaustively	Exhaustive small group rewiring	$\Omega(2^n)$

In the table, n is the number of inputs to the circuit and m is the number of gates. Currently known BDD-based and most circuit-based methods can detect only a fraction of all symmetries in some cases, while the method based on graph automorphism (this work) can detect all symmetries exhaustively. Additionally, the symmetry-detection techniques in this work find all phase-shift symmetries as well as composite (hybrid) symmetries that simultaneously involve both permutations and phase-shifts. In contrast, existing literature on functional symmetries does not consider such composite symmetries.

A comparison of BDD-based symmetry detection [100], circuit-based symmetry detection [32, 109, 134], Boolean-matching-based symmetry detection [28] and the method proposed in this paper is summarized in Table 7.1.

7.1.3 Graph-Automorphism Algorithms

Our symmetry-detection method is based on efficient graph-automorphism algorithms, which have recently been improved by Darga et al. [58, 59]. Their symmetry detector Saucy finds all symmetries of a given colored undirected graph. To this end, consider an undirected graph G with n vertices, and let $V = \{0, ..., n-1\}$. Each vertex in G is labeled with a unique value in V. A permutation on V is a bijection $\pi : V \to V$. An automorphism of G is a permutation π of the labels assigned to vertices in G such that $\pi(G) = G$; we say that π is a structure-preserving mapping or symmetry. The set of all such valid relabellings is called the automorphism group of G. A coloring is a restriction on the permutation of vertices – only vertices in the same color can map to each other. Given G, possibly with colored vertices, Saucy produces symmetry generators that form a compact description of all symmetries.

7.1.4 Post-Placement Rewiring

Rewiring based on symmetries can be used to optimize circuit characteristics. Some rewiring examples are illustrated in Figure 7.1(a), (b). For the discussion in this chapter the goal is to reduce wirelength, and swapping symmetric input and output pins accomplishes this.

C.-W. Chang et al. [32] use the symmetry-detection technique described in Section 7.1.2 to optimize delay, power and reliability. In general, symmetry detection in their work is done opportunistically rather than exhaustively. Experimental results show that their approach can achieve these goals effectively using the symmetries detected. However, they cannot find the rewiring opportunities in Figure 7.1(a), (b) because their symmetry-detection technique lacks the ability to detect output and phase-shift symmetries.

Another type of rewiring is based on the addition and removal of wires. Three major techniques are used to determine the wires that can be reconnected. The first one uses reasoning based on ATPG such as REWIRE [47], RAMFIRE [33] and the work by Jiang et al. [76], which tries to add a redundant wire that makes the target wire redundant so that it can be removed. The second class of techniques is graph-based; one example is GBAW [138], which uses pre-defined graph representation of subcircuits and relies on pattern matching to replace wires. The third technique uses SPFDs [55] and is based on don't-cares. Although these techniques are potentially more powerful than symmetry-based rewiring because they allow more aggressive layout changes, they are also less stable and do not support post-silicon metal fix.

7.2 Exhaustive Search for Functional Symmetries

The symmetry-detection method presented in our work can find all input, output, multi-variable and phase-shift symmetries including composite (hybrid) symmetries. It relies on symmetry detection of graphs, thus the original Boolean function must be converted to a graph first. After that, it solves the graph-automorphism (symmetry detection) problem on this graph, and then the symmetries found are converted to symmetries of the original Boolean function. Our main contribution is the mapping from a Boolean function to a graph, and showing how to use it to find symmetries of the Boolean function. These techniques are described in detail in this section.

7.2.1 Problem Mapping

To reduce functional symmetry detection to the graph-automorphism problem, we represent Boolean functions by graphs as described below:

1 Each input and its complement are represented by two vertices in the graph, and there is an edge between them to maintain Boolean consistency (i.e., $x \leftrightarrow y$ and $x' \leftrightarrow y'$ must happen simultaneously). These vertices are called *input vertices*. Outputs are handled similarly, and the vertices are called *output vertices*.

2 Each minterm and maxterm of the Boolean function is represented by a *term vertex*. We introduce an edge connecting every minterm vertex to the output and an edge connecting every maxterm vertex to the complement of the output. We also introduce an edge between every term vertex and every input vertex or its complement, depending on whether that input is 1 or 0 in the term.

3 Since inputs and outputs are bipartite-permutable, all input vertices have one color, and all outputs vertices have another color. All term vertices use yet another color.

The idea behind this construction is that if an input vertex gets permuted with another input vertex, the term vertices connected to them will also need to be permuted. However, the edges between term vertices and output vertices restrict such permutations to the following cases: (1) the permutation of term vertices does not affect the connections to output vertices, which means the outputs are unchanged; and (2) permuting term vertices may also require permuting output vertices, thus capturing output symmetries. A proof of correctness is given in Section 7.2.2.

Figure 7.2(a) shows the truth table of function $z = x \oplus y$, and Figure 7.2(b) illustrates our construction for the function. In general, vertex indices are assigned as follows. For n inputs and m outputs, the ith input is represented

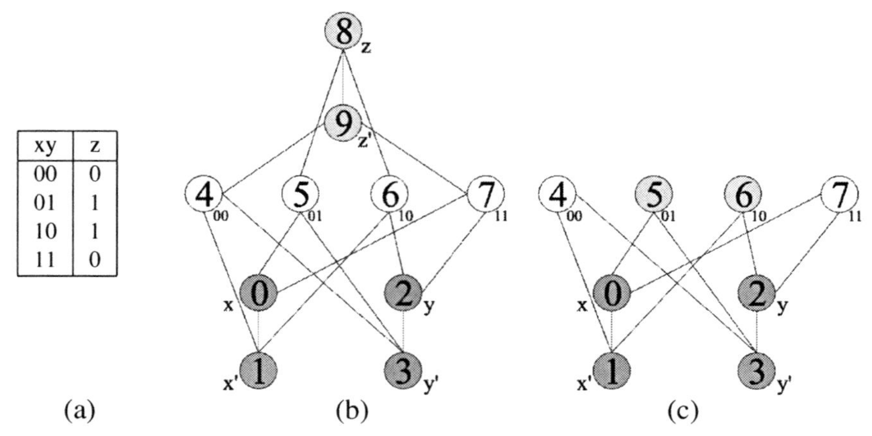

xy	z
00	0
01	1
10	1
11	0

Figure 7.2. Representing the 2-input XOR function by (**a**) the truth table, (**b**) the full graph, and (**c**) the simplified graph for faster symmetry detection.

by vertex $2i$, while the complementary vertex has index $2i + 1$. There are 2^n terms, and the ith term is indexed by $2n+i$. Similarly, the ith output is indexed by $2n + 2^n + 2i$, while its complement is indexed by $2n + 2^n + 2i + 1$.

The symmetry detector Saucy [58, 59] used in this work typically runs faster when the graph is smaller and contains more colors. Therefore if output symmetries do not need to be detected, a simplified graph with reduced number of vertices can be used. It is constructed similarly to the full graph, but without output vertices and potentially with more vertex colors. We define an *output pattern* as a set of output vertices in the full graph that are connected to a given term vertex. Further, term vertices with different output patterns shall be colored differently. Figure 7.2(c) illustrates the simplified graph for the two-input XOR function.

All the minterms and maxterms of the Boolean function are used in the graph because we focus on fully-specified Boolean functions. Since all the terms are used, and there are 2^n terms for an n-input function, the time and space complexity of our algorithm is $\Omega(2^n)$.

7.2.2 Proof of Correctness

We first prove the correctness of the simplified graph construction proposed in the previous section. Our proofs below are presented as a series of numbered steps.

1 First, we need to prove that there is a one-to-one mapping between the function and its graph. This mapping can be defined following the graph construction in Section 7.2.1. The inverse mapping (from a graph to a function) is also given in the section.

2 Second, we need to prove that there is a one-to-one mapping between symmetries of the function and automorphisms of the graph.

   (a) First, we want to show that a symmetry of the function is an automorphism of the graph. A symmetry of the function is a permutation of the function's inputs that do not change the function's outputs, and permutation in inputs corresponds to reevaluation of the outputs of that the term. Since the inputs are symmetric, no output will be changed by the permutation, and the color of the term vertices in the corresponding graph will remain the same. Therefore it is also an automorphism of the graph.

   (b) Next we want to show that an automorphism of the graph is a symmetry of the function. Since there is an edge between the input and its complement, mapping one input vertex, say x, to another vertex, say y, will cause x's complement map to y's complement, so Boolean consistency is preserved. Since an input vertex connect to all the term vertices that contain it, swaps between two input vertices will cause all the term vertices that connect to them being swapped according to the following rule: suppose that input vertex x swaps with input vertex y, then all term vertices that connect to both x and y will also be swapped because there is an edge between the term vertex and both x and y. Since a swap between term vertices is legal only if they have the same color, it means all automorphisms detected in the graph will not map a term vertex to another color. And since the color of the term represents an output pattern in the Boolean function, it means the outputs of the Boolean function will not be changed. Therefore an automorphism of the graph maps to an input symmetry of the Boolean function.

3 From Steps 1 and 2, there is a one-to-one mapping between the function and its graph, and a one-to-one mapping between the symmetries of the function and the automorphisms of the graph. Therefore the symmetry-detection method for the simplified graph is correct.

Next, the correctness of the original graph is proved below. The relationship between terms and inputs are described in the previous proof. Therefore the proof here focuses on the relationship between terms and outputs. There are three possible situations: input symmetries that do not affect the outputs, input symmetries that affect the outputs, and output symmetries that are independent of the inputs.

1 Input symmetries that do not affect the output: the way term vertices connect to output vertices represent an output pattern. If two term vertices have exactly the same outputs, then they will connect to the same output

vertices; otherwise they will connect to at least one different output vertex. Mapping a term vertex to another term vertex which has different output pattern is invalid (except for the situation described in 2) because at least one output vertex they connect to is different, therefore the connections to output vertices behave the same as coloring in the previous proof.

2 Input symmetries that affect the output: if all terms that connect to an output pattern can be mapped to all terms connecting to another output pattern, then the output vertices corresponding to the two patterns can also be swapped because the terms that the outputs connect to will not change after the mapping. In the mean time, the input vertices that connect to the swapped minterms will also be swapped, which represent a symmetry involving both inputs and outputs.

3 Output symmetries that are independent of the inputs: if two sets of output vertices connect to exactly the same term vertices, then the output vertices in the two sets can be swapped, which represent output symmetries. In this case, no term swapping is involved, so the inputs are unaffected.

7.2.3 Generating Symmetries from Symmetry Generators

The symmetry detector Saucy returns symmetry generators. To produce symmetries that can be used for rewiring, we design a $symmetry\_generation$ algorithm, which is shown in Figure 7.3. In the algorithm, $generators$ is a set which contains all the symmetry generators returned by Saucy, and three sets of symmetries are used. They are old_sym, cur_sym and new_sym. Initially, cur_sym contains the identity symmetry (i.e., a symmetry that maps to itself), and both old_sym and new_sym are empty. The algorithm loops until cur_sym is empty, meaning that all the symmetries have been generated; or $count$ is larger than 1000, meaning that 1000 symmetries have been generated. As a result, at most 1000 symmetries will be tried for a set of symmetry generators to limit the complexity of rewiring. When the loop finishes, old_sym will contain all the symmetries generated using the generators.

7.2.4 Discussion

Compared with other symmetry-detection methods, the symmetry detector described in this chapter has the following advantages: (1) it can detect all possible input and output symmetries of a function, including multi-variable, higher-order and phase-shift symmetries; and (2) symmetry generators are used to represent the symmetries, which make the relationship between input and output symmetries very clear. These characteristics make the use of the symmetries easier than other methods that enumerate all symmetry pairs.

While evaluating our algorithm, we observed that Saucy is more efficient when there are few or no symmetries in the graph; in contrast, it takes more

```
Function symmetry_generation(generators)
1   do
2     foreach sym ∈ cur_sym
3       foreach gen ∈ generators
4         for i = 1 to 2 do
5           nsym = (i == 1)?gen × sym : sym × gen;
6           if (!nsym ∈ (old_sym ∪ cur_sym ∪ new_sym))
7             new_sym = new_sym ∪ nsym;
8             count = count + 1;
9     old_sym = old_sym ∪ cur_sym;
10    cur_sym = new_sym;
11    new_sym.clear();
12  while (!cur_sym.empty() and count < 1000);
13  return old_sym;
```

Figure 7.3. Our symmetry generation algorithm.

time when there are many symmetries. For example, the runtime of a randomly chosen 16-input function is 0.11 sec because random functions typically have no symmetries. However, it takes 9.42 sec to detect all symmetries of the 16-input XOR function. Runtimes for 18 inputs are 0.59 and 92.39 sec, respectively.

7.3 Post-Placement Rewiring

This section describes a permutative rewiring technique that uses symmetries of extracted subcircuits to reduce wirelength. Implementation insights and further discussions are also given.

7.3.1 Permutative Rewiring

After placement, symmetries can be used to rewire the netlist to reduce the wirelength. This is achieved by reconnecting pins according to the symmetries found in subcircuits, and these subcircuits are extracted as follows.

1. We represent the netlist by a hypergraph, where cells are represented by nodes and nets are represented by hyper-edges.

2. For each node in the hypergraph, we perform Breadth-First Search (BFS) starting from the node, and use the first n nodes traversed as subcircuits.

3. Similarly, we perform Depth-First Search (DFS) and extract subcircuits using the first m nodes.

In our implementation, we perform BFS extraction 4 times with n from 1 to 4, and DFS twice with m from 3 to 4. This process is capable of extracting various subcircuits suitable for rewiring. In addition to logically connected cells, min-cut placers such as Capo [3, 24] produce a hierarchical collection of *placement bins* (buckets) that contain physically adjacent cells, and these bins are also suitable for rewiring. Currently, we also use subcircuits composed of cells in every half-bin and full-bin in our rewiring. After subcircuits are extracted, we perform symmetry detection on these subcircuits. Next, we reconnect the wires to the inputs and outputs of these subcircuits according to the detected symmetries in order to optimize wirelength.

The reason why multiple passes with different sizes of subcircuits are used is that some symmetries in small subcircuits cannot be detected in larger subcircuits. For example, in Figure 7.4, if the subcircuit contains all the gates, only symmetries between x, y, z and w can be detected, and the rewiring opportunity for p and q will be lost. By using multiple passes for symmetry detection, more symmetries can be extracted from the circuit.

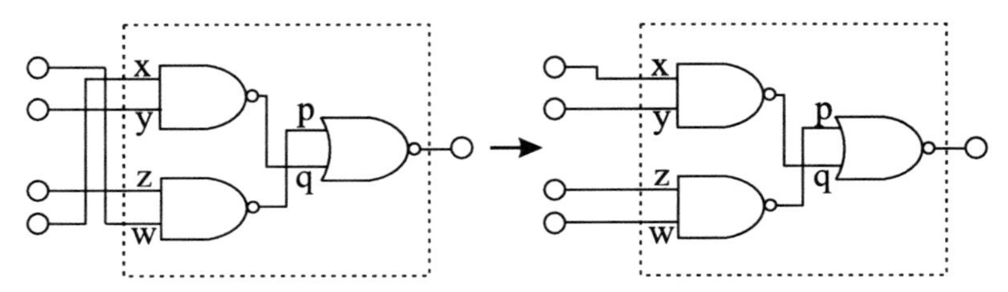

Figure 7.4. Rewiring opportunities for p and q cannot be detected by only considering the subcircuit shown in this figure. To rewire p and q, a subcircuit with p and q as inputs must be extracted.

The rewiring algorithm can be easily extended to utilize phase-shift symmetry: if the wirelength is shorter after the necessary inverters are inserted or removed, then the circuit is rewired. It can also be used to reduce the delay on critical paths.

7.3.2 Implementation Insights

During implementation, we observed that for subcircuits with a small number of inputs and outputs, it is more efficient to detect symmetries by enumerating all possible permutations using bit operations on the truth table. That is because the required permutations can be implemented with just a few lines of C++ code, making this technique much faster than building the graph for Saucy. We call this algorithm *naive symmetry detection*. To further reduce its runtime, we limit the algorithm to detect first-order symmetries only. In our implementation, naive symmetry detection is used on subcircuits with number of inputs less than 11 and number of outputs less than 3. Experimental results show that the runtime can be reduced by more than half with almost no loss in quality, which is because the lost rewiring opportunities can be recovered in larger subcircuits where Saucy-based symmetry detection is used.

7.3.3 Discussion

Our rewiring techniques described so far use permutational symmetries. Here we describe two applications of phase-shift symmetries.

1 The ability to handle phase-shift symmetries may reduce interconnect by enabling permutational symmetries, as the MUX example in Figure 7.1(b) shows.

2 Phase-shift symmetries support metal fix of logic errors involving only inversions of signals: by reconnecting certain wires, signals may be inverted.

Compared with other rewiring techniques, the advantages of our techniques include the following:

- Our rewiring techniques preserve placement, therefore the effects of any change are immediately measurable. As a result, our methods are safe and can be applied with every flow. In other words, their application can only improve the optimization objective and never worsens it. This characteristic is especially desirable in highly-optimized circuits because changes in placements may create cell overlaps, and the legalization process to remove these overlaps may affect multiple gates, leading to a deterioration of the optimization goal.

- Our techniques support post-silicon metal fix, which allows reuse of transistor masks and can significantly reduce respin cost.

- The correctness of our optimizations can be verified easily using combinational equivalence checking.

- Our techniques can optimize a broad variety of objectives, as long as the objectives can be evaluated incrementally.

The limitations of our rewiring techniques include:

- The performance varies with each benchmark, depending on the number of symmetries that exist in a design. Therefore improvement is not guaranteed.

- When optimizing wirelength, the ratio of improvement tends to reduce when designs get larger. Since permutative rewiring is a local optimization, it cannot shorten global nets.

7.4 Experimental Results

Our implementation was written in C++, and the testcases were selected from ITC99, ISCAS and MCNC benchmarks. To better reflect modern VLSI circuits, we chose the largest testcases from each benchmark suite, and added several small and medium ones for completeness. Our experiments used the min-cut placer Capo. The platform used was Fedora 2 Linux on a Pentium-4 workstation running at 2.26 GHz with 512 M RAM.

We converted every testcase from BLIF to the Bookshelf placement format (*.nodes* and *.nets* files) using the converter provided in [25, 156]. We report two different types of experimental results in this section, including the number of symmetries detected and rewiring. A flow chart of our experiments on symmetry detection and rewiring is given in Figure 7.5.

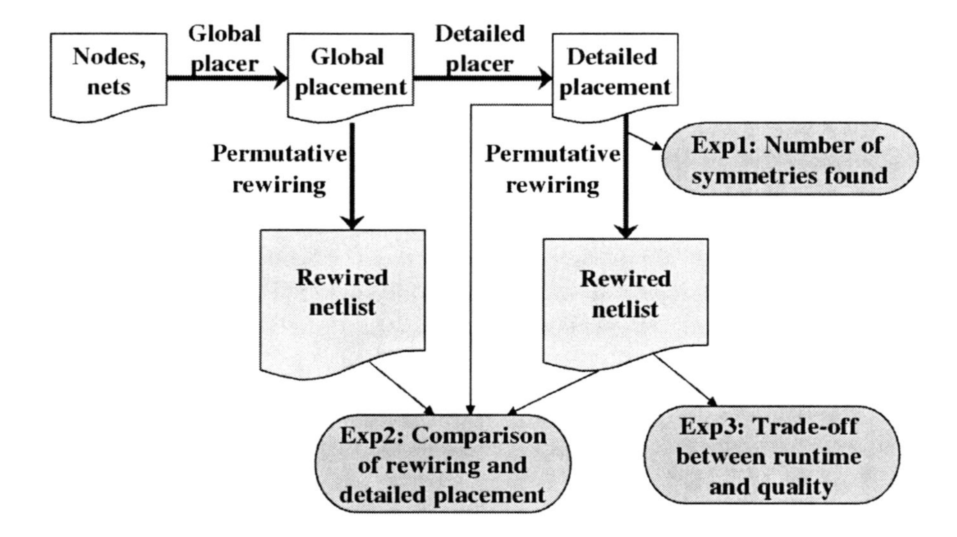

Figure 7.5. Flow chart of our symmetry detection and rewiring experiments.

7.4.1 Symmetries Detected

The first experiment evaluates the symmetries found in the benchmarks, and the results are summarized in Table 7.2. In the table, "number of subcircuits" is the number of subcircuits extracted from the benchmark for symmetry detection. "Input" is the number of subcircuits which contain input symmetries, and "phase-shift input" is the number of subcircuits that contain phase-shift input symmetries. "Output" and "phase-shift output" are used in a similar way. "Input and output" are subcircuits that contain symmetries involving both inputs and outputs. The number of symmetries found in the circuits can be used to predict the probability of finding rewiring opportunities: at least 66% of the subcircuits contain permutational input symmetries and are suitable for rewiring. It can also be observed that although output symmetries do

Table 7.2. Number of symmetries found in benchmark circuits.

Benchmark	Number of subcircuits	Symmetries				
		Input	Phase-shift input	Output	Phase-shift output	Input and output
ALU2	876	855	120	249	126	211
ALU4	15933	15924	242	1245	243	1244
B02	143	130	18	22	15	21
B10	1117	1015	160	201	137	170
B17	198544	190814	23789	32388	17559	24474
C5315	20498	19331	9114	5196	4490	4145
C7552	28866	26626	12243	7540	6477	5895
DALU	16665	15506	6632	3272	2550	2852
I10	14670	14165	4298	3710	2929	2516
S38417	141241	126508	75642	64973	59319	61504
S38584	122110	117084	55966	35632	29661	33655
Average	100%	94%	28%	23%	18%	20%

Row "Average" shows the average percentages of subcircuits that contain a specific symmetry type. For example, the number in the last row of the third column means 94% of the subcircuits contain at least one input symmetry.

not happen as often as input symmetries, their number is not negligible and rewiring techniques should also take output symmetries into consideration.

7.4.2 Rewiring

In the rewiring experiments, wirelength reduction was calculated against the original wirelength after placement using half-perimeter wirelength. The second experiment compares the wirelength reduction gained from rewiring and detailed placement. It also compares the wirelength reduction of rewiring before and after detailed placement. These results are summarized in Tables 7.3 and 7.4, respectively. The maximum number of inputs allowed for symmetry detection was 16 in this experiment. From Table 7.3, it is found that our method can effectively reduce wirelength by approximately 3.7%, which is comparable to the improvement due to detailed-placement.

Table 7.4 shows that the wirelength reduction is a little bit smaller when rewiring is used after detailed placement, suggesting that some rewiring opportunities interfere with optimization from detailed placement. For example, detailed placement performs flipping of cells, which may interfere with permutative rewiring if the inputs of the cell are symmetric. However, the difference is very small, showing that wirelength reduction from rewiring is mostly independent of detailed placement.

Table 7.3. Wirelength reduction and runtime comparisons between rewiring, detailed placement and global placement.

Benchmark	Wirelength	Wirelength reduction		Runtime (seconds)		
		Rewiring (%)	Detailed placement (%)	Rewiring	Detailed placement	Global placement
ALU2	5403.29	3.21	8.98	2.6	0.2	3.6
ALU4	35491.38	9.02	3.54	15.2	3.0	27.2
B02	142.90	8.29	0.00	2.8	0.4	0.1
B10	1548.28	5.04	3.89	7.2	0.1	1.0
B17	367223.20	2.92	2.28	350.6	32.6	206.2
C5315	30894.06	1.76	1.52	17.39	3.0	3.2
C7552	39226.30	1.71	1.57	23.8	4.0	2.8
DALU	20488.84	2.79	3.46	13.2	2.6	2.6
I10	50613.84	2.11	2.05	15.6	2.6	29.0
S38417	129313.20	2.01	2.05	180.8	22.2	17.2
S38584	174232.80	2.51	2.27	157.8	20.6	46.0
Average	77689	3.70	2.87	30.8	8.3	71.5

Table 7.4. The impact of rewiring before and after detailed placement.

Benchmark	Wirelength reduction		Runtime (seconds)	
	Before detailed placement(%)	After detailed placement(%)	Before detailed placement	After detailed placement
ALU2	3.49	3.21	3.4	3.6
ALU4	9.38	9.02	27.2	27.2
B02	8.29	8.29	0.2	0.2
B10	4.78	5.04	0.8	1.0
B17	3.00	2.92	199.6	206.2
C5315	1.71	1.76	3.6	3.2
C7552	1.82	1.71	2.6	2.8
DALU	2.90	2.19	2.8	2.6
I10	2.05	2.11	29.2	29.0
S38417	2.04	2.01	18.0	17.2
S38584	2.50	2.51	46.2	46.0
Average	3.82	3.70	30.3	30.8

The third experiment evaluates the relationship between the number of inputs allowed in symmetry detection, wirelength reduction and runtime. In order to show the true performance of Saucy-based symmetry detection, the use of naive symmetry detection was turned off in this experiment. Since our symmetry-detection method is most efficient with small number of inputs, this relationship represents the trade-off between performance and runtime. Empirical results are shown in Table 7.5, where the numbers are averages of all the benchmarks. These results indicate that the longer the rewiring program runs, the better the reduction will be. However, most improvement occurs with small number of inputs and can be achieved quickly. In addition, recent follow-up work by Chai et al. [30] showed how to simplify the graphs that represent logic functions in order to speed up symmetry detection. Their techniques can make our symmetry detector run faster and thus further improve the rewiring quality given the same amount of time.

Table 7.5. The impact of the number of inputs allowed in symmetry detection on performance and runtime.

Number of inputs allowed	Runtime (seconds)	Wirelength reduction(%)
2	2.90	1.06
4	4.30	2.58
6	7.07	3.12
8	14.98	3.50
10	28.03	3.63
12	41.34	3.72
14	59.85	3.66
16	82.30	3.68

We also applied our rewiring techniques to the OpenCores suite [154] in the IWLS'05 benchmarks [161], and we performed routing to measure the wirelength reduction for routed wires. The results show that our pre-routing optimizations transform into post-routing wirelength reduction effectively. Furthermore, we observe that via counts can also be reduced by our optimizations. These results show that our rewiring techniques are effective in reducing wirelength and number of vias, and they can both reduce manufacturing defects and improve yield. Reducing via count is especially important in deep submicron era because vias are a major cause of manufacturing faults. Detailed results are reported in [37].

7.5 Summary

In this chapter we presented a new symmetry-detection methodology and applied it to post-placement rewiring. Compared with other symmetry-detection techniques, our method identifies more symmetries, including multi-variable permutational and phase-shift symmetries for both inputs and outputs. This is important in circuit rewiring because more detected symmetries create more rewiring opportunities.

Our experimental results on common circuit benchmarks show that the wirelength reduction is comparable and orthogonal to the reduction provided by detailed placement – the reduction achieved by our method performed before and after detailed placement is similar. This shows that our rewiring method is very effective, and it should be performed after detailed placement for the best results. When applied, we observe an average of 3.7% wirelength reduction for the experimental benchmarks evaluated.

In summary, the rewiring technique we presented has the following advantages: (1) it does not alter the placement of any standard cells, therefore no cell overlaps are created and improvements from changes can be evaluated reliably; (2) it can be applied to a variety of existing design flows; (3) it can optimize a broad variety of objectives, such as delay and power, as long as they can be evaluated incrementally; and (4) it can easily adapt to other symmetry detectors, such as the detectors proposed by Chai et al. [29, 30]. On the other hand, our technique has some limitations: (1) its performance depends on the specific design being optimized and there is no guarantee of wirelength reduction; and (2) the improvement tends to decrease with larger designs, similar to what has been observed from detailed placement.

PART III

FOGCLEAR COMPONENTS

Chapter 8

BUG TRACE MINIMIZATION

Finding the cause of a bug can be one of the most time-consuming activities in design verification. This is particularly true in the case of bugs discovered in the context of a random simulation-based methodology, where bug traces, or counterexamples, may be several hundred thousand cycles long. In this chapter we describe Butramin, a bug trace minimizer. Butramin considers a bug trace produced by a random simulator or a semi-formal verification software and produces an equivalent trace of shorter length. Butramin applies a range of minimization techniques, deploying both simulation-based and formal methods, with the objective of producing highly reduced traces that still expose the original bug. Our experiments show that in most cases Butramin is able to reduce traces to a small fraction of their initial sizes, in terms of cycle length and signals involved. The minimized traces can greatly facilitate bug analysis. In addition, they can also be used to reduce regression runtime.

8.1 Background and Previous Work

Research on minimizing property counterexamples or, more generally, bug traces, has been pursued both in the context of hardware and software verification. Before discussing these techniques, we first give some preliminary background.

8.1.1 Anatomy of a Bug Trace

A *bug state* is an undesirable state that exposes a bug in the design. Depending on the nature of the bug, it can be exposed by a unique state (a specific bug configuration) or any one of several states (a general bug configuration), as shown in Figure 8.1. In the figure, suppose that the x-axis represents one state machine called FSM-X and the y-axis represents another machine called

FSM-Y. If a bug occurs only when a specific state in FSM-X and a specific state in FSM-Y appear simultaneously, then the bug configuration will be a very specific single point. On the other hand, if the bug is only related to a specific state in FSM-X but it is independent of FSM-Y, then the bug configuration will be all states on the vertical line intersecting the one state in FSM-X. In this case, the bug configuration is very broad.

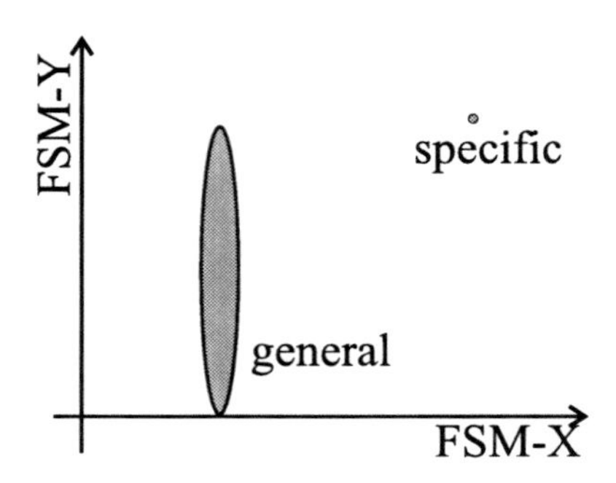

Figure 8.1. An illustration of two types of bugs, based on whether one or many states expose a given bug. The x-axis represents FSM-X and the y-axis represents FSM-Y. A specific bug configuration contains only one state, while a general bug configuration contains many states.

Given a sequential circuit and an initial state, a *bug trace* is a sequence of test vectors that exposes a bug, i.e., causes the circuit to assume one of the bug states. The *length* of the trace is the number of cycles from the initial state to the bug state, and an *input event* is a change of an input signal at a specific clock cycle of the trace. One input event is considered to affect only a single input bit. An *input variable assignment* is a value assignment to an input signal at a specific cycle. The term *input variable assignment* is used in the literature when traces are modeled as sequences of symbolic variable assignments at the design's inputs. The number of input variable assignments in a trace is the product of the number of cycles and the number of inputs. A *checker signal* is a signal used to detect a violation of a property. In other words, if the signal changes to a specific value, then the property monitored by the checker is violated, and a bug is found. The objective of *bug trace minimization* is to reduce the number of input events and cycles in a trace, while still detecting the checker violation.

EXAMPLE 8.1 *Consider a circuit with three inputs a, b and c, initially set to zero. Suppose that a bug trace is available where a and c are assigned to* 1 *at cycle* 1. *At cycle* 2, *c is changed to* 0 *and it is changed back to* 1 *at cycle* 3, *after which a checker detects a violation. In this situation we count four input events, twelve input variable assignments, and three cycles for our bug trace. The example trace is illustrated in Figure 8.2.*

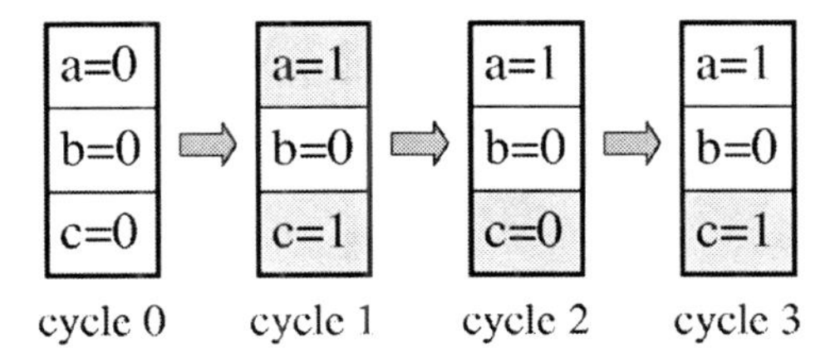

Figure 8.2. A bug trace example. The boxes represent input variable assignments to the circuit at each cycle, shaded boxes represent input events. This trace has three cycles, four input events and twelve input variable assignments.

Another view of a bug trace is a path in the state space from the initial state to the bug state, as shown in Figure 8.3. By construction, formal methods can often find the minimal length bug trace as shown in the dotted line. Therefore we focus our minimization on semi-formal and constrained-random traces only. However, if Butramin is applied to a trace obtained with a formal technique, it may still be possible to reduce the number of input events and variable assignments.

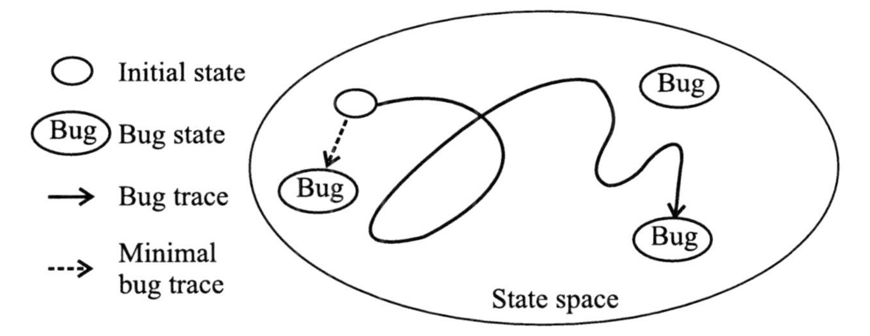

Figure 8.3. Another view of a bug trace. Three bug states are shown. Formal methods often find the minimal length bug trace, while semi-formal and constrained-random techniques often generate longer traces.

8.1.2 Known Techniques in Hardware Verification

Traditionally, a counterexample generated by BMC reports the input variable assignments for each clock cycle and for each input of the design. However, it is possible, and common, that only a portion of these assignments are required to falsify the property. Several techniques that attempt to minimize the trace complexity have been recently proposed, for instance, Ravi et al. [113]. To this end they propose two techniques: *brute-force lifting (BFL)*, which attempts to eliminate one variable assignment at a time, and an improved variant that eliminates variables in such a way so as to highlight the primary events that led to the property falsification. The basic idea of BFL is to consider the *free variables* of the bug trace, that is, all input variable assignments in every cycle. For each free variable v, BFL constructs a SAT instance SAT(v) to determine if v can prevent the counterexample. If that is not the case, then v is irrelevant to the counterexample and can be eliminated. Because this technique minimizes BMC-derived traces, its focus is only on reducing the number of assignments to the circuit's input signals. Moreover, each single assignment elimination requires solving a distinct SAT problem, which may be computationally difficult. More recent work in [121] further improves the performance of BFL by attempting the elimination of sets of variables simultaneously. Our technique for removing individual variable assignments is similar to BFL as it seeks to remove an assignment by evaluating a trace obtained with the opposite assignment. However, we apply this technique to longer traces obtained with semi-formal methods and we perform testing via resimulation.

Another technique applied to model checking solutions is by Gastin et al. [65]. Here the counterexample is converted to a *Büchi automaton* and a depth-first search algorithm is used to find a minimal bug trace. Minimization of counterexamples is also addressed in [77], where the distinction between control and data signals is exploited in attempting to eliminate data signals first from the counterexample.

All of these techniques focus on reducing the number of input variable assignments to disprove the property. Because the counterexample is obtained through a formal model checker, the number of cycles in the bug trace is minimal by construction. Butramin's approach considers a more general context where bug traces can be generated by simulation or semi-formal verification software, attacking much more complex designs than BMC-based techniques. Therefore, (1) traces are in general orders of magnitude longer than the ones generated by BMC; and (2) there is much potential for reducing the trace in terms of number of clock cycles, as our experimental results indicate. On the downside, the use of simulation-based techniques does not guarantee that the results obtained are of minimal length. As the experimental results in Section 8.5 indicate, however, our heuristics provide good results for most benchmarks.

Aside from minimization of bug traces generated using formal methods, techniques that generate traces by random simulation have also been explored in the context of hardware verification. One such technique is by Chen et al. [50] and proceeds in two phases. The first phase identifies all the distinct states of the counterexample trace. The second phase represents the trace as a state graph: it applies one step of forward state traversal [57] to each of the individual states and adds transition edges to the graph based on it. Dijkstra's shortest path algorithm is applied to the final graph obtained. This approach, while very effective in minimizing the trace length (the number of clock cycles in the trace), (1) does not consider elimination of input variable assignments, and (2) makes heavy use of formal state-traversal techniques, which are expensive computationally and can usually be applied only to small-size designs, as indicated also by the experimental results in [50].

8.1.3 Techniques in Software Verification

The problem of trace minimization has been a focus of research also in the software verification domain. Software bug traces are characterized by involving a very large number of variables and very long sequences of instructions. The delta debugging algorithm [69] is fairly popular in the software world. It simplifies a complex software trace by extracting the portion of the trace that is relevant to exposing the bug. Their approach is based exclusively on resimulation-based exploration and it attacks the problem by partitioning the trace (which in this case is a sequence of instructions) and checking if any of the components can still expose the bug. The algorithm was able to greatly reduce bug traces in Mozilla, a popular web browser. A recent contribution that draws upon counterexamples found by model checking is by Groce et al. [68]. Their solution focuses on minimizing a trace with respect to the primitive constructs available in the language used to describe the hardware or software system and on trying to highlight the causes of the error in the counterexample, so as to produce a simplified trace that is more understandable by a software designer.

8.2 Analysis of Bug Traces

In this section, we analyze the characteristics of bug traces generated using random simulation, pointing out the origins of redundancy in these traces and propose how redundancy can be removed. In general, redundancy exists because some portions of the bug trace may be unrelated to the bug, there may be loops or shortcuts in the bug trace, or there may be an alternative and shorter path to the bug. Two examples are given below to illustrate the idea, while the following subsections provide a detailed analysis.

EXAMPLE 8.2 *Intel's first-generation Pentium processor included a bug in the floating-point unit which affected the FDIV instruction. This bug occurred when FDIV was used with a specific set of operands. If there had been a checker testing for the correctness of the FDIV operation during the simulation-based verification of the processor, it is very probable that a bug trace exposing this problem may be many cycles long. However, only a small portion of the random program would have been useful to expose the FDIV bug, while the majority of other instructions can be eliminated. The redundancy of the bug trace comes from the cycles spent testing other portions of the design, which are unrelated to the flawed unit and can thus be removed.*

EXAMPLE 8.3 *Suppose that the design under test is a FIFO unit, and a bug occurs every time the FIFO is full. Also assume that there is a pseudo-random bug trace containing both read and write operations until the trace reaches the "FIFO full" state. Obviously, cycles that read data from the FIFO can be removed because they create state transitions that bring the trace away from the bug configuration instead of closer to it.*

8.2.1 Making Traces Shorter

In general, a trace can be made shorter if any of the following situations arise: (a) it contains loops; (b) there are alternative paths (shortcuts) between two design states; or (c) there is another state which exposes the same bug and can be reached earlier.

The first situation is depicted schematically in Figure 8.4. In random simulation, a state may be visited more than once, and such repetitive states will form loops in the bug trace. Identifying such loops and removing them can reduce the length of the bug trace.

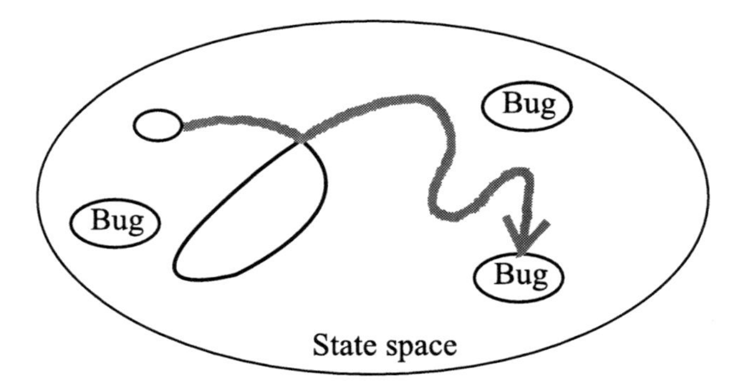

Figure 8.4. A bug trace may contain sequential loops, which can be eliminated to obtain an equivalent but more compact trace.

In the second case, there may be a shortcut between two states as indicated by arrow 1 in Figure 8.5, which means an alternative path may exist from a state to another state using fewer cycles. Such situations may arise in random traces frequently because constrained-random simulation often selects transitions arbitrarily and it is possible that longer paths are generated in place of shorter ones.

The third condition occurs when multiple design states exist that expose the same bug, and some of them can be reached in fewer steps compared to the original one, as shown by arrows marked "2" in Figure 8.5. If a path to those states can be found, it is possible to replace the original one.

A heuristic approach that can be easily devised to search for alternative shorter traces is based on generating perturbations on a given trace. A bug trace

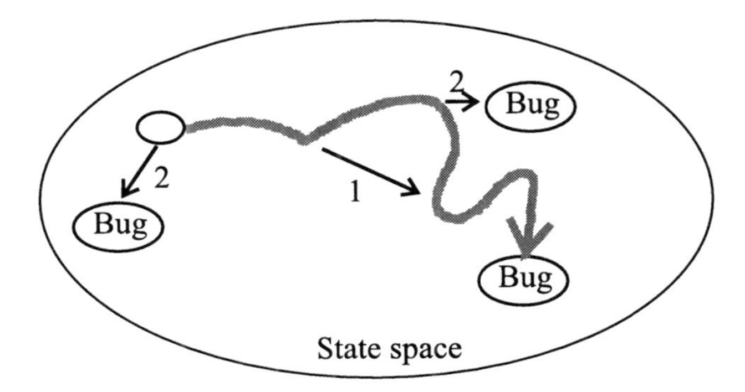

Figure 8.5. Arrow 1 shows a shortcut between two states on the bug trace. Arrows marked "2" show paths to easier-to-reach bug states in the same bug configuration (that violate the same property).

can be perturbed locally or globally to find shortcuts or a path to an alternative bug state. In a *local perturbation*, cycles or input events are added or removed from an original trace. As mentioned previously, random simulation selects state transitions in a pseudo-random fashion. By local perturbation, alternative transitions can be explored and shorter paths to a trace state or to another state exposing the bug may be found. In a *global perturbation*, a completely new trace is generated, and the trace can be used to replace the original one if it is shorter. One reason why perturbation has the potential to work effectively on random traces is that a pseudo-random search tends to do a large amount of local exploration, compared to a formal trace that progresses directly to a bug. Because of this, opportunities of shortcuts within a trace abound.

8.2.2 Making Traces Simpler

After all redundant cycles are removed, many input events may still be left. For example, if a circuit has 100 inputs and a bug trace is 100 cycles long, there are 10,000 input variable assignments in the trace. However, not all assignments are relevant to expose the bug. Moreover, redundant events increase the complexity of interpreting the trace in the debugging phase. Therefore it is important to identify and remove such redundancy.

We envision two ways of simplifying the input assignments in a trace: by removing input events and by eliminating assignments that are not essential to reach our goal. In this latter approach, input assignments can be marked as essential or not, based on their impact in exposing the bug. By removing nonessential input variable assignments, the analysis of the bug trace during debugging can be made much simpler. For example, a trace with two input events will be much easier to analyze than a trace with 10,000 input events.

8.3 Proposed Techniques

Based on our analysis, we propose several techniques to minimize a bug trace. In this section we first provide an overview of these techniques, and then we discuss each technique in detail.

1 *Single-cycle elimination* shortens a bug trace by resimulating a variant of the trace which includes fewer simulation cycles.

2 *Alternative path to bug* is exploited by detecting when changes made on a trace produce an alternative, shorter path to the bug.

3 *State skip* identifies all the unique state configurations in a trace. If the same state occurs more than once, it indicates the presence of a loop between two states, and the trace can be reduced.

4 *BMC-based refinement* attempts to further reduce the trace length by searching locally for shorter paths between two trace states.

In addition, we propose the following techniques to simplify traces:

1 *Input-event elimination* attempts to eliminate input events, by resimulating trace variants which involve fewer input events.

2 *Essential variable identification* uses three-value simulation to distinguish essential variable assignments from nonessential ones, and marks the nonessentials with "X".

3 Indirectly, all cycle removal techniques may also remove redundant input events.

A bug trace can be perturbed by either adding or removing cycles or input events. However, trying all possibilities is infeasible. Since the purpose of minimization is to reduce the number of cycles and input events, we only use removal in the hope to find shorter and simpler traces. Our techniques are applied in the following order: Butramin first tries to shorten a trace by removing certain clock cycles and simulating such trace variants, after which it tries to reduce the number of input events. While analyzing each perturbed trace, the two techniques of alternative path to bug and state skip monitor for loops and shorter paths. Once these techniques run out of steam, Butramin applies a series of BMC refinements. The BMC search is localized so that we never generate complex SAT instances for SAT solving, which could become the bottleneck of Butramin. If our SAT solver times out on some BMC instances, we simply ignore such instances and potential trace reductions since we do not necessarily aim for the shortest traces.

8.3.1 Single-Cycle Elimination

Single-cycle elimination is an aggressive but efficient way to reduce the length and the number of input events in a bug trace. It tentatively removes a whole cycle from the bug trace and checks if the bug is still exposed by the new trace through resimulation, in which case the new shorter trace replaces the old one. This procedure is applied iteratively on each cycle in the trace, starting from cycle 1 and progressing to the end of the trace. The reason we start from the first simulation cycle is that perturbing early stages of a trace has a better chance to explore states far away from the original trace. The later a removal the less the opportunity to visit states far away from the original trace.

EXAMPLE 8.4 *Consider the trace of Example 8.1. During the first step, single-cycle elimination attempts to remove cycle* 1. *If the new trace still exposes the bug, we obtain a shorter bug trace which is only two cycles long and has two input events, as shown in Figure 8.6. Note that it is possible that some input events become redundant because of cycle elimination, as it is the case in this example for the event on signal* c *at cycle* 2. *This is because the previous transition on* c *was at cycle 1, which has now been removed. After events which have become redundant are eliminated, single-cycle elimination can be applied to cycle* 2 *and* 3, *iteratively.*

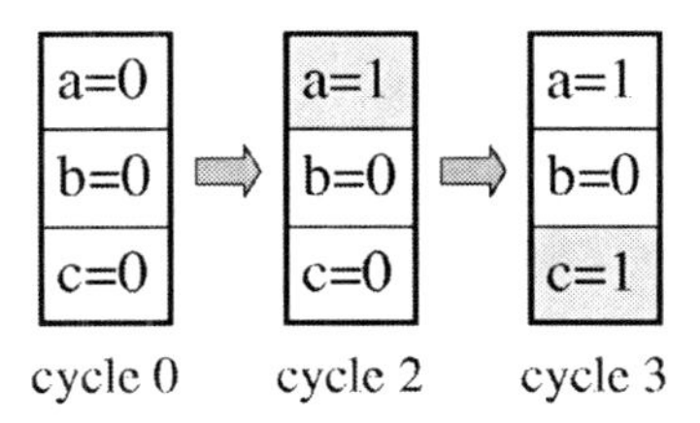

Figure 8.6. Single-cycle elimination attempts to remove individual trace cycles, generating reduced traces which still expose the bug. This example shows a reduced trace where cycle 1 has been removed.

To reduce Butramin's runtime, we extend single-cycle elimination to work with several cycles at once. When three consecutive cycles are eliminated one by one, Butramin will try to eliminate pairs of consecutive cycles. If that succeeds, the next attempt will consider twice as many cycles. If it fails, the number of cycles considered at once will be halved. This *adaptive cycle elimination* technique can dynamically extend its "window size" to quickly eliminate large sequences of cycles when this is likely, but will roll back to single-cycle removal otherwise.

Note that, when dependency exists between blocks of cycles, removing a single cycle at a time may invalidate the bug trace. For example, removing any cycle within a PCI-X transaction will almost always corrupt the transaction, rendering the bug trace useless. This problem can be addressed by removing whole transactions instead of cycles. With some extra inputs from the user to help identify transaction boundaries, Butramin can be easily adapted to handle transaction-based traces.

8.3.2 Input-Event Elimination

Input-event elimination is the basic technique to remove input events from a trace. It tentatively generates a variant trace where one input event is replaced by the complementary value assignment. If the variant trace still exposes the bug, the input event can be removed. In addition, the event immediately following on the same signal becomes redundant and can be removed as well.

EXAMPLE 8.5 *Consider once again the trace of Example 8.1. The result after elimination of input event c at cycle* 1 *is shown in Figure 8.7. Note that the input event on signal c at cycle* 2 *becomes redundant and it is also eliminated.*

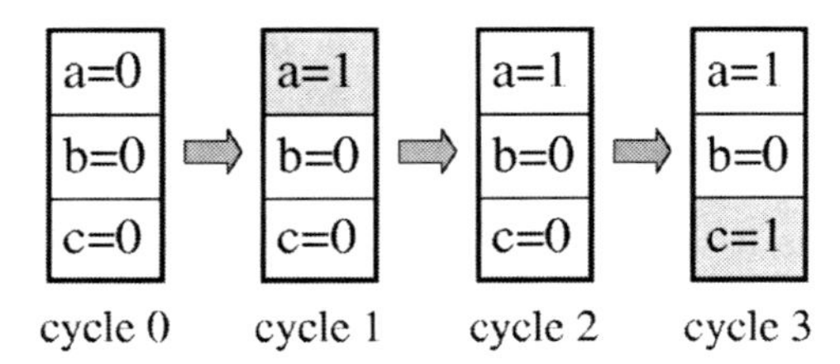

Figure 8.7. Input-event elimination removes pairs of events. In the example, the input events on signal c at cycle 1 and 2 are removed.

8.3.3 Alternative Path to Bug

An alternative path to bug occurs when a variant trace reaches a state that is different from the final state of the trace, but it also exposes the same bug. The alternative state must obviously be reached in fewer simulation steps than in the original trace. As shown in Figure 8.8, if state s_{j2}, reached at time t_2 by the variant trace (shown at the bottom) exposes the bug, the new variant trace replaces the original one.

8.3.4 State Skip

The state skip rule is useful when two identical states exist in a bug trace. This happens when there is a sequential loop in the trace or when, during the simulation of a tentative variant trace, an alternative (and shorter) path to a state in the original trace is found. Consider the example shown in Figure 8.9: if states s_{j_2} and s_{i_4} are identical, then a new, more compact trace can be generated by appending the portion from step t_5 and on of the original trace, to the prefix extracted from the variant trace up to and including step t_2. This technique identifies all reoccurring states in a trace and removes cycles

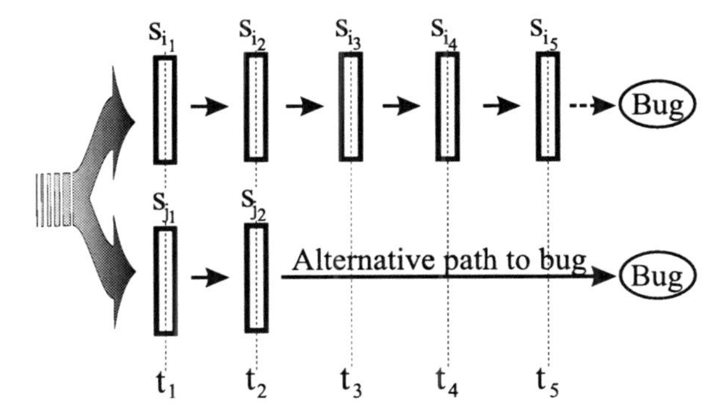

Figure 8.8. Alternative path to bug: the variant trace at the bottom hits the bug at step t_2. The new trace replaces the old one, and simulation is stopped.

between them, guaranteeing that all the states in the final minimized trace are unique. States are hashed for fast look-up so that state skip does not become a bottleneck in execution.

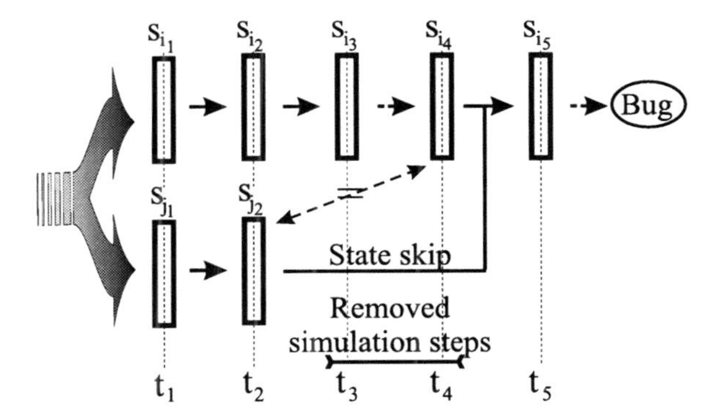

Figure 8.9. State skip: if state $s_{j_2} = s_{i_4}$, cycles t_3 and t_4 can be removed, obtaining a new trace which includes the sequence "... s_{j_1}, s_{j_2}, s_{i_5}, ...".

8.3.5 Essential Variable Identification

We found that, after applying our minimization techniques, bug traces are usually much shorter. However, many input variable assignments may still be part of the trace, and their relevance in exposing the bug may vary – some may be essential, while others are not. Butramin includes an "X-mode" feature for filtering out irrelevant input variable assignments, where input variable assignments are classified as essential or not, based on a 3-value (0/1/X) simulation analysis. To implement this technique, two bits are used to encode each signal value, and each input assignment at each cycle is assigned in turn the value X: if the X input propagates to the checker's output and an X is sampled on the checker's output signal, then the input is marked essential, and the original input assignment is kept. Otherwise, the input assignment is deemed irrelevant for the purpose of exposing the bug. The set of input assignments that are marked irrelevant contribute to simplify the debugging activity, since a verification engineer does not need to take them into consideration when studying the cause of the system's incorrect behavior. We present experimental results indicating that this analysis is capable of providing substantial simplifications to the signals involved in an already reduced bug trace.

Note, finally, that our simplification technique, which relies on 3-value simulation, is over-conservative, flagging irrelevant input assignments as essential. Consider, for instance, the simulation of a multiplexer where we propagated an X value to the select input and a 1 value to both data inputs. A 3-valued logic simulator would generate X at the output of the simulator; however, for our purposes, the correct value should have been 1, since we consider X to mean "don't-care". If more accuracy is desired for this analysis, a hybrid logic/symbolic simulator can be used instead [82, 136].

Alternatively, essential variable identification could be performed using a BMC-based technique with a pseudo-Boolean SAT solver, for instance [61, 166]. Such solvers satisfy a given SAT formula with the smallest possible number of assigned variables (maximal number of don't-cares). Aside from these solvers, even mainstream Boolean SAT solvers can be specialized to do this, as suggested in [113]. Since assignments in the SAT solution correspond to input variable assignments in the bug trace, those input variable assignments are obviously essential. Essential variable identification naturally follows by marking all other input variable assignments as irrelevant. A similar idea has been deployed also by Lu et al. [95] to find a three-valued solution which minimizes the number of assignments to state variables.

8.3.6 BMC-Based Refinement

This technique can be used after simulation-based minimization to further reduce the length of a bug trace. Because of state skip, after applying simulation-based minimization, no two states in a trace will be the same. However, shorter

```
1  Select two states s_i and s_j, k cycles apart;
2  for l = 1 to k − 1 do
3     C = circuit unrolled l times;
4     Transform C into a Boolean formula CNF_c;
5     I=CNF_c ∧ CNF_{s_i} ∧ CNF_{s_j};
6     if (I is satisfiable)
7        return (shortcut s_i → s_j, l steps);
```

Figure 8.10. BMC-based shortcut detection algorithm.

paths between any pair of states may still exist. We propose here an approach based on model checking to find such paths. The algorithm, also outlined in Figure 8.10, considers two states, say s_i and s_j, which are k cycles apart in the trace and attempts to find the shortest path connecting them. This path can then be found by unrolling the circuit from 1 to $k-1$ times, asserting s_i and s_j as the initial and final states, and attempting to satisfy the corresponding Boolean formula. If we refer to the CNF formula of the unrolled circuit as CNF_c, then $CNF_c \wedge CNF_{s_i} \wedge CNF_{s_j}$ is the Boolean formula to be satisfied. If a SAT solver can find a solution, then we have a shortcut connecting s_i to s_j. Note that the SAT instances generated by our algorithm are simplified by the fact that CNF_{s_i} and CNF_{s_j} are equivalent to a partial satisfying assignment for the instance. An example is given in Figure 8.11.

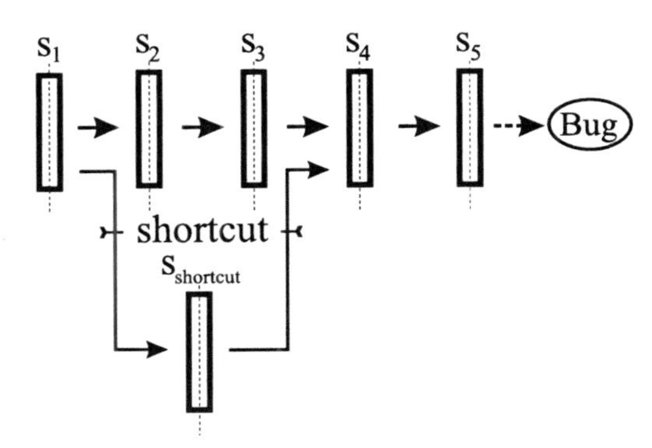

Figure 8.11. BMC-based refinement finds a shortcut between states S_1 and S_4, reducing the overall trace length by one cycle.

The algorithm described in Figure 8.10 is applied iteratively on each pair of states that are k steps apart in the bug trace, and using varying values for k from 2 to m, where m is selected experimentally so that the SAT instance can be solved efficiently. We then build an explicit directed graph using the shortcuts found by the BMC-based refinement and construct the final shorter path from the initial state to the bug state. Figure 8.12 shows an example of such graph. Each vertex in the graph represents a state in the starting trace, edges between vertices represent the existence of a path between the corresponding states, and the edge's weight is the number of cycles needed to go from the source state to the sink. Initially, there is an edge between each two consecutive vertices, and the weight labels are 1. Edges are added between vertices when shortcuts are found between the corresponding states, and they are labeled with the number of cycles used in the shortcut. A single-source shortest path algorithm for directed acyclic graphs is then used to find a shorter path from the initial to the bug state. While some of the shortcuts discovered by BMC may be incompatible because of the partial constraints in CNF_{s_i}

and CNF_{s_j}, the algorithm we describe selects an optimal set of compatible shortcuts within the selected window size m.

Although simulation-based techniques are effective, they are heuristic in nature and may miss local optimization opportunities. BMC-based refinement has the potential to improve on local optimizations by performing short-range optimal cycle elimination.

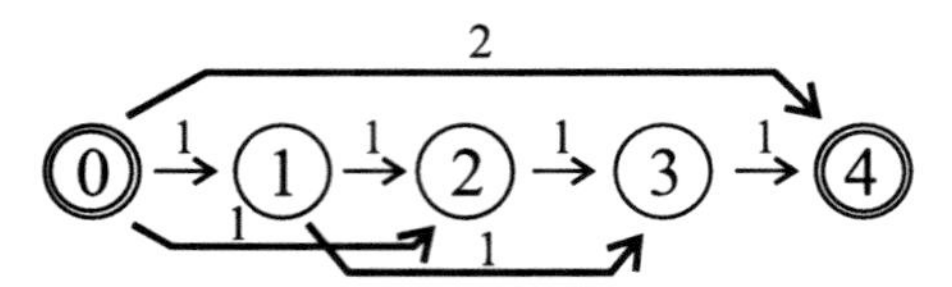

Figure 8.12. A shortest-path algorithm is used to find the shortest sequence from the initial state to the bug state. The edges are labeled by the number of cycles needed to go from the source vertex to the sink. The shortest path from states 0 to 4 in the figure uses 2 cycles.

8.4 Implementation Insights

We built a prototype implementation of the techniques described in the previous section to evaluate Butramin's performance and trace reduction capability on a range of digital designs. Our implementation strives to simplify a trace as much as possible, while providing good performance at the same time. This section discusses some of the insights we gained while constructing a Butramin's prototype.

8.4.1 System Architecture

The architecture of Butramin consists of three primary components: a driver program, commercial logic simulation software, and a SAT solver. The driver program is responsible for (1) reading the bug trace, (2) interfacing to the simulation tool and SAT solver for the evaluation of the compressed variant traces, and (3) finding simplifications introduced in the previous sections. The logic simulation software is responsible for simulating test vectors from the driver program, notifying the system if the trace reaches the bug under study, and communicating back to the driver each visited state during the simulation. BMC-based minimization was implemented using MiniSat [61] that analyzes the SAT instances generated by converting the unrolled circuits to CNF form using a CNF generator. The system architecture is shown in Figure 8.13.

8.4.2 Algorithmic Analysis and Performance Optimizations

In the worst case scenario, the complexity of our simulation-based techniques is quadratic in the length of the trace under evaluation, and linear in the size of the primary input signals of the design. In fact, consider an m-cycle

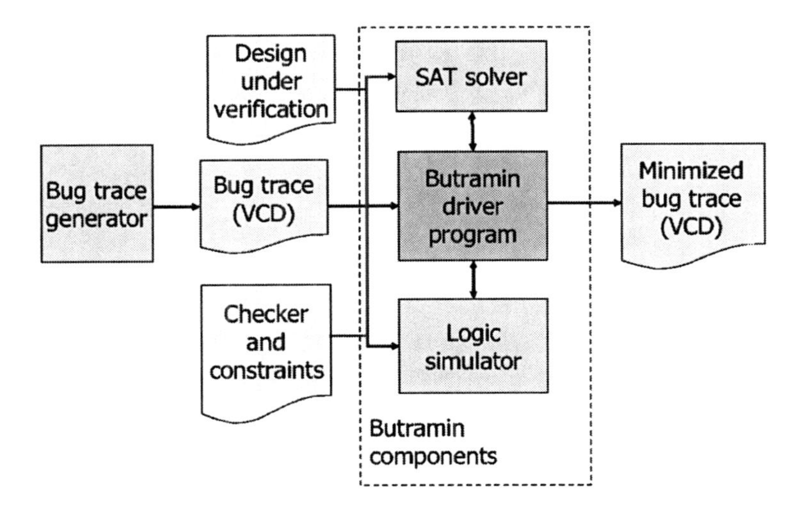

Figure 8.13. Butramin system architecture.

long bug trace driving an n-input design. The worst case complexity for our cycle-elimination technique is $O(m^2)$, where the one of the input-event elimination technique is $O(n \times m^2)$. All the other simulation-based techniques have simpler complexity or are independent of the size of the trace or design. In order to improve the runtime of Butramin, we developed an extra optimization as described below. Experimental results show that the worst case situation did not occur due to our optimization, adaptive cycle elimination and the nature of practical benchmarks.

The optimization focuses on identifying all multiple occurrences of a state so that we can identify when the simulation of a variant trace falls into the original trace, and then we can avoid simulating the last portion of the variant. To achieve this, we hash all states visited by a trace and tag them with the clock cycle in which they occur. During the simulation of variant traces we noted that, in some special conditions, we can improve the performance of Butramin by reducing the simulation required: after the time when the original and the variant traces differ, if a variant state matches a state in the original trace tagged by the same clock cycle, then we can terminate the variant simulation and still guarantee that the variant trace will hit the bug. In other words, simulation can be terminated early because the result of applying the same test vectors after the matched state will not change. We call this an *early exit.* As illustrated in Figure 8.14, early exit points allow the simulation to terminate immediately. Often simulations can also be terminated early by state skip optimization because the destination state is already in the trace database. Experimental results show that this optimization is crucial to the efficiency of simulation-based minimization techniques.

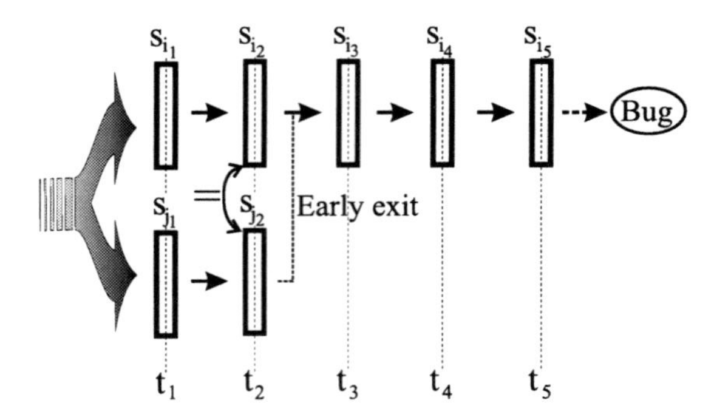

Figure 8.14. Early exit. If the current state s_{j_2} matches a state s_{i_2} from the original trace, we can guarantee that the bug will eventually be hit. Therefore, simulation can be terminated earlier.

8.4.3 Use Model

To run Butramin, the user must supply four inputs: (1) the design under test, (2) a bug trace, (3) the property that was falsified by the trace, and (4) an optional set of constraints on the design's input signals. Traces are represented as Value Change Dump (VCD) files, a common compact format that includes all top-level input events. Similarly, the minimized bug traces are output as VCD files.

Removing input events from the bug trace during trace minimization may generate illegal input sequences, which in turn could erroneously falsify a property or make the trace useless. For example, removing the reset event from a bug trace may lead the design into an erroneous state, generating a spurious trace which does not reflect a possible legal activity of the design under verification, even if the simulation of such trace does expose the original design flaw. Consequently, when testing sub-components of a design with constrained inputs, it becomes necessary to validate the input sequences generated during trace minimization. There are several ways to achieve this goal. One technique is to mark required inputs so that Butramin does not attempt to remove the corresponding events from the trace. This approach is a viable solution to handle, for instance, reset and the clock signals. For complex sets of constraints, it is possible to convert them into an equivalent circuit block connected to the original design, such as the techniques described in the work by Yuan et al. [144]. This extra circuit block takes random input assignments and converts them into a set of legal assignments which satisfy all the required environment constraints. We deployed the former approach for simple situations, and we adapted the latter to the context of our solution for benchmarks with more complex environments. Specifically, since Butramin starts already with a valid input trace which it attempts to simplify, we wrote our constraints as a set of monitors which observe each input sequence to the design. If the monitors flag an illegal transition during simulation, the entire "candidate trace" is deemed invalid and removed from consideration. For BMC-based refinement, these environmental constraints are synthesized and included as additional constraints to the problem instance. Note, however, that this limits BMC-based techniques to be applied to designs whose environmental constraints are synthesizable. On the other hand, this requirement is lifted for the simulation-based minimization techniques. From our experimental results, we observe that most minimization is contributed by simulation-based techniques, which renders this requirement optional for most practical benchmarks.

We also developed an alternative use model to apply Butramin to reducing regression runtime. In this context, the approach is slightly different since the goal now is to obtain shorter traces that achieve the same functional coverage as their longer counterpart. To support this, coverage points are encoded by properties: each of them is "violated" only when the corresponding point is covered by the trace. Butramin can then be configured to generate traces that violate all of the properties so that the same coverage is maintained.

8.5 Experimental Results

We evaluated Butramin by minimizing traces generated by a range of commercial verification tools: a constrained-random simulator, semi-formal verification software, and again a semi-formal tool where we specified to use extra effort in generating compact traces. We considered nine benchmark designs from OpenCores (FPU), ISCAS89 (S15850, S38584), ITC99 (B15), IWLS2005 (VGALCD), picoJava (picoJava, ICU), as well as two internally developed benchmarks (MULT, DES), whose characteristics are reported in Table 8.1. We developed assertions to be falsified when not already available with the design, and we inserted bugs in the design that falsify the assertions. Table 8.2 describes assertions and bugs inserted. For ICU and picoJava, no bugs were injected but the constraints for random simulation were relaxed. The checker for VGALCD is a correct duplicate of the original design (which we modified to contain one design error), hence the circuit size we worked with is twice as the one reported in Table 8.1. Finally, experiments were conducted on a Sun Blade 1500 (1 GHz UltraSPARC IIIi) workstation running Solaris 9.

8.5.1 Simulation-Based Experiments

Our first set of experiments attempts to minimize traces generated by running a semi-formal commercial verification tool with the checkers specified, and subsequently applying only the simulation-based minimization techniques of Butramin, described in Sections 8.3.1, 8.3.2, 8.3.3, and 8.3.4. We were not able to complete the generation of traces with the semi-formal verification tool for VGALCD, therefore we only report results related to constrained-random

Table 8.1. Characteristics of benchmarks.

Benchmark	Inputs	Flip-flops	Gates	Description
S38584	41	1426	20681	Unknown
S15850	77	534	10306	Unknown
MULT	257	1280	130164	Wallace tree multiplier
DES	97	13248	49183	DES algorithm
B15	38	449	8886	Portion of 80386
FPU	72	761	7247	Floating point unit
ICU	30	62	506	PicoJava instruction cache unit
picoJava	53	14637	24773	PicoJava full design
VGALCD	56	17505	106547	VGA/LCD controller

The benchmark setup for VGALCD involves duplicating this design and modifying one connection in one of the copies. Butramin then minimizes the trace exposing the difference. It follows that the size of the benchmark we work with is actually twice as the one reported for this design.

Table 8.2. Bugs injected and assertions for trace generation.

Circuit	Bug injected	Assertion used
S38584	None	Output signals forced to a specific value
S15850	None	Output signals forced to a specific value
MULT	AND gate changed with XOR	Compute the correct output value
DES	Complemented output	Timing between receive_valid, output_ready and transmit_valid
B15	None	Coverage of a partial design state
FPU	divide_on_zero conditionally complemented	Assert divide_on_zero when divisor=0
ICU	Constraints relaxed	Buffer-full condition
picoJava	Constraints relaxed	Assert SMU's spill and fill
VGALCD	Circuit duplicated with one wire changed in one copy	Outputs mismatch condition

traces for this benchmark. Table 8.3 shows the absolute values of cycles and input events left in each trace and the overall runtime of Butramin using only simulation-based techniques. Figures 8.15 and 8.16 show the percentages of cycles and input events removed from the original bug trace using different techniques. Note that for all benchmarks we are able to remove the majority of cycles and input events.

Table 8.3. Cycles and input events removed by simulation-based techniques of Butramin on traces generated by semi-formal verification.

Circuit	Cycles			Input events			Runtime
	Original	Remain	Removed (%)	Original	Remain	Removed (%)	(seconds)
S38584	13	8	38.46	255	2	99.22	19
S15850	59	1	98.31	2300	3	99.87	5
MULT	345	4	98.84	43843	2	99.99	35
DES	198	154	22.22	3293	3	99.91	254
B15	25015	11	99.96	450026	15	99.99	57
FPU	53711	5	99.99	1756431	17	99.99	27
ICU	6994	3	99.96	62740	3	99.99	5
picoJava	30016	10	99.97	675485	11	99.99	3359

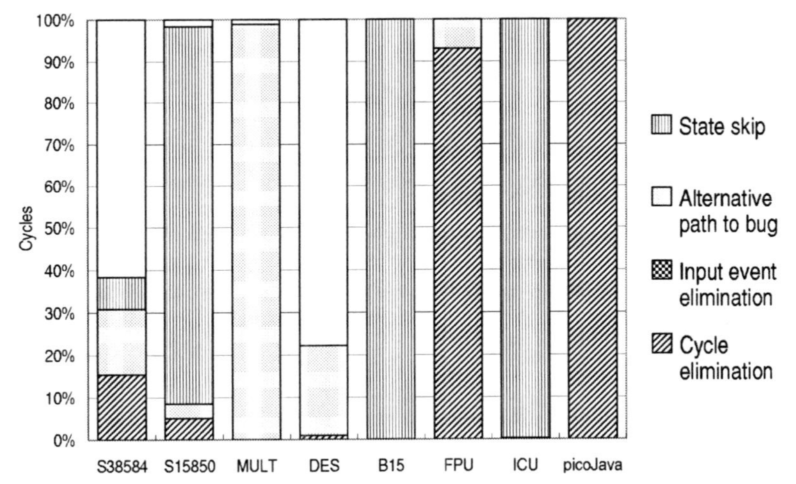

Figure 8.15. Percentage of cycles removed using different simulation-based techniques. For benchmarks like B15 and ICU, state skip is the most effective technique because they contain small numbers of state variables and state repetition is more likely to occur. For large benchmarks with long traces like FPU and picoJava, cycle elimination is the most effective technique.

With reference to Figures 8.15 and 8.16, we observe that the contribution of different minimization techniques varies among benchmarks. For example, almost all the cycles and input events are removed by cycle elimination in FPU and picoJava. On the other hand, state skip removes more than half of the cycles and input events in B15 and ICU. This difference can be attributed to the nature of the benchmark: if there are fewer state variables in the design, state skip is more likely to occur. In general, state skip has more opportunities to provide trace reductions in designs that are control-heavy, such as ICU, compared to designs that are datapath-heavy, such as FPU and picoJava. Although input-event elimination does not remove cycles, it has great impact in eliminating input events for some benchmarks, such as S38584. Overall, we found that all these techniques are important to compact different types of bug traces.

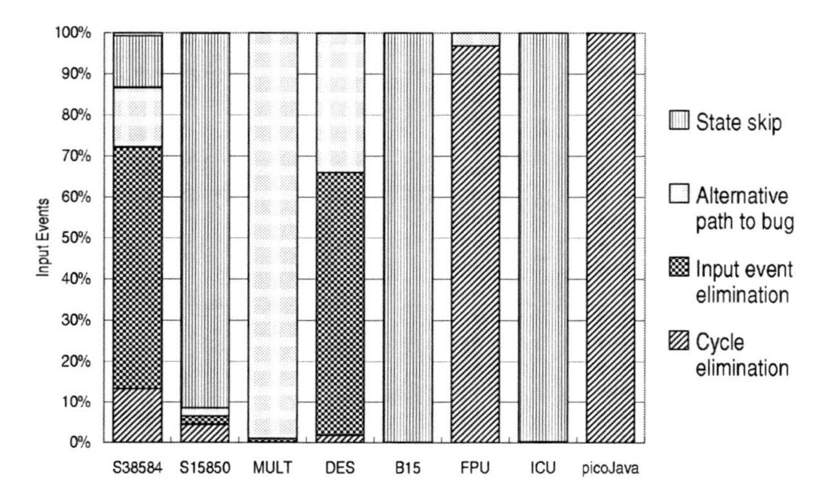

Figure 8.16. Number of input events eliminated with simulation-based techniques. The distributions are similar to cycle elimination because removing cycles also removes input events. However, input-event elimination works the most effectively for some benchmarks like S38584 and DES, showing that some redundant input events can only be removed by this technique.

Our second set of experiments applies Butramin to a new set of traces, also generated by a semi-formal tool, but this time we configured the software to dedicate extra effort in generating short traces, by allowing more time to be spent on the formal analysis of the checker. Similar to Table 8.3 discussed earlier, Table 8.4 reports the results obtained by applying the simulation-based minimization techniques of Butramin to these traces. We still find that Butramin has a high impact in compacting these traces, even if, generally speaking, they present less redundancy, since they are closer to be minimal. Note in particular, that the longer the traces, the greater the benefit from the application of Butramin. Even if the overall impact is reduced, we still observe a 61% reduction in the number of cycles and 91% in input events, on average.

The third set of experiments evaluated traces generated by constrained-random simulation. Results are summarized in Table 8.5. As expected, Butramin produced the most impact on this set of traces, since they tend to include a large amount of redundant behavior. The average reduction is 99% in terms of cycles and input events.

Table 8.4. Cycles and input events removed by simulation-based techniques of Butramin on traces generated by a compact-mode semi-formal verification tool.

Circuit	Cycles			Input events			Runtime
	Original	Remain	Removed (%)	Original	Remain	Removed (%)	(seconds)
S38584	13	8	38.46	255	2	99.22	21
S15850	17	1	94.12	559	56	89.98	4
MULT	6	4	33.33	660	2	99.70	34
DES	296	17	94.26	3425	3	99.91	17
B15	27	11	59.26	546	5	99.08	6
FPU	23	5	78.26	800	17	97.88	1
ICU	19	14	26.32	142	80	43.66	1
picoJava	26	10	61.54	681	11	98.38	39

Table 8.5. Cycles and input events removed by simulation-based methods of Butramin on traces generated by constrained-random simulation.

Circuit	Cycles			Input events			Runtime
	Original	Remain	Removed (%)	Original	Remain	Removed (%)	(seconds)
S38584	1003	8	99.20	19047	2	99.99	16
S15850	2001	1	99.95	77344	3	99.99	2
MULT	1003	4	99.60	128199	2	99.99	34
DES	25196	154	99.39	666098	3	99.99	255
B15	148510	10	99.99	2675459	9	99.99	395
FPU	1046188	5	99.99	36125365	17	99.99	723
ICU	31992	3	99.99	287729	3	99.99	5
picoJava	99026	10	99.99	2227599	16	99.99	5125
VGALCD	36595	4	99.99	1554616	19	99.99	28027

8.5.2 Performance Analysis

Table 8.6 compares Butramin's runtime with and without different optimization techniques. The traces are generated using semi-formal methods in this comparison. The execution runs that exceeded 40,000 seconds were timed-out (T/O in the table). The runtime comparison shows that early exit and state skip have great impacts on the execution time: early exit can stop resimulation early, and state skip may reduce the length of a trace by many cycles at a time. Although these two techniques require extra memory, the reduction in runtime shows they are worthwhile. In ICU, state skip occurred 4 times, removing 6977 cycles, which resulted in a very short runtime. The comparison also shows that adaptive cycle elimination iscapable of reducing minimization

Table 8.6. Impact of the various simulation-based techniques on Butramin's runtime.

Benchmark	Runtime(seconds)		
	[1]: cycle elimination+ input-event elimination	[2]: [1]+state skip+ early exit	[3]: [2]+adaptive cycle elimination
S38584	21	19	19
S15850	11	5	5
MULT	48	43	35
DES	274	256	254
B15	T/O	58	57
FPU	T/O	235	27
ICU	8129	5	5
picoJava	T/O	T/O	3359
Average	1697	66	64

Benchmarks that exceeded the time limit (40,000s) are not included in the average. Each of the runtime columns reports the runtime using only a subset of our techniques: the first cycle elimination and input-event elimination. The second includes in addition early exit and state skip, and the third adds also adaptive cycle elimination.

time significantly. This technique is especially beneficial for long bug traces, such as FPU and picoJava.

A comparison of Butramin's impact and runtime on the three sets of traces is summarized in Figure 8.17. The result shows that Butramin can effectively reduce all three types of bug traces in a reasonable amount of time. Note, in addition, that in some cases the minimization of a trace generated by random simulation takes similar or less time than applying Butramin to a trace generated by a compact-mode semi-formal tool, even if the initial trace is much longer. That is the case for S38584 or S15850. We explain this effect by the nature of the bug traces: traces generated by random simulation tend to visit states that are easily reachable, therefore states are likely to be repetitive, and state skip occurs more frequently, leading to a shorter minimization time. On the other hand, states visited in a compact-mode generated trace are more frequently produced by formal engines and can be highly specific, making state skip a rare event. The cases of FPU and picoJava are relevant in this context: here state skips do not occur, and the minimization time is highly related to the original trace length. They also demonstrate the benefits of Butramin in various verification methodologies.

8.5.3 Essential Variable Identification

We also applied the technique from Section 8.3.5 to identify essential variables from the minimized traces we generated. Table 8.7 shows that after this technique is applied, many input variable assignments are marked nonessential,

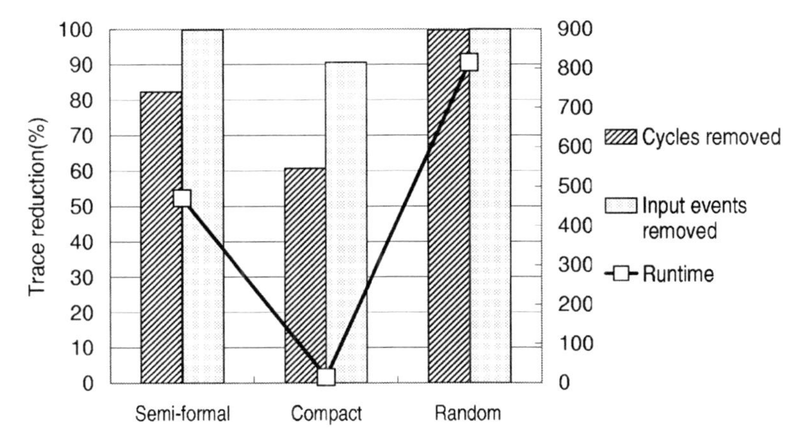

Figure 8.17. Comparison of Butramin's impact when applied to traces generated in three different modes. The graph shows the fraction of cycles and input events eliminated and the average runtime.

further simplifying the trace. Note that the comparison is now between input variable assignments, not input events. Since all nonessential input variable assignments are simulated with X, the simulation will propagate X values to many internal signals as well. As a result, it will be easier to understand the impact of essential variable assignments on violated properties.

Table 8.7. Essential variable assignments identified in X-mode.

Circuit	Input variables	Essential variables
S38584	320	2
S15850	76	2
MULT	1024	1019
DES	14748	2
B15	407	45
FPU	355	94
ICU	87	21
picoJava	520	374

The table compares the number of input variable assignments in the minimized traces with the number of assignments classified essential. All the remaining assignments are nonessential and can be replaced by X values in simulation. The initial traces were generated by a semi-formal verification tool.

8.5.4 Generation of High-Coverage Traces

In order to evaluate the effectiveness of Butramin on reducing regression runtime, we selected three benchmarks, DES, FPU and VGALCD, as our

Table 8.8. Cycles and input events removed by simulation-based methods of Butramin on traces that violate multiple properties.

Circuit	#Pro-	Cycles			Input events			Runtime
	perties	Original	Remain	Removed (%)	Original	Remain	Removed (%)	(seconds)
DES	2	25196	184	99.27	666098	17	99.99	549
FPU	3	1046188	9	99.99	36125365	264	99.99	580
VGALCD	3	36595	5	99.98	1554616	22	99.99	25660

multi-property benchmarks. The original properties in the previous experiments were preserved, and the same traces generated by constrained-random simulation were used. In addition, we included a few extra properties, so that our original traces would expose them before reaching their last simulation step. Those extra properties specify a certain partial states to be visited or a certain output signals to be asserted. Butramin is then configured to produce minimized traces that violate all properties. The results are summarized in Table 8.8. Compared with Table 8.5, it can be observed that in order to cover extra properties, the length of the minimized traces are now longer. However, Butramin continues to be effective for these multi-property traces. We also found that the order of property violations is preserved before and after minimization, suggesting that Butramin minimizes segments of bug traces individually. From an algorithmic complexity point of view, minimizing a multi-property trace is similar to minimizing many single-property traces with different initial states.

While the original traces of FPU and VGALCD require 20–30 minutes to be simulated, post-Butramin traces are short enough to be simulated in just a few seconds. The benefits of adding the minimized trace to a regression suite, instead of the original one, are obvious.

8.5.5 BMC-Based Experiments

We applied our BMC-based technique to traces already minimized using simulation-based methods to evaluate the potential for further minimization. For VGALCD, we report only data related to the minimization of random trace since semi-formal traces are not available. The results are summarized in Table 8.9, where *Orig* is the original number of cycles in the trace, and *Removed* is the number of cycles removed by this method. We used a maximum window of 10 cycles ($m = 10$). The main observation that can be made is that simulation-based techniques are very effective in minimizing bug traces. In fact, only in two cases, ICU and B15, our BMC-based technique was able to extract additional minimization opportunities. Potentially, we could repeat the application of simulation-based techniques and BMC-based methods until convergence, when no additional minimization can be extracted.

Table 8.9. Cycles removed by the BMC-based method.

Circuit	Semi-formal			Compact-trace			Constrained-random		
	Orig	Removed	Time	Orig	Removed	Time	Orig	Removed	Time
S38584	8	0	55 s	8	0	55 s	8	0	55 s
S15850	1	0	2 s	1	0	2 s	1	0	2 s
MULT	4	0	20 s	4	0	20 s	4	0	20 s
DES	154	0	23 h3m	17	0	357 s	154	0	23 h3m
B15	**11**	**1**	**121 s**	**11**	**1**	**121 s**	10	0	97 s
FPU	5	0	5 s	5	0	5 s	5	0	5 s
ICU	**3**	**1**	**1 s**	**14**	**2**	**1 s**	**3**	**1**	**1 s**
picoJava	10	0	70 s	10	0	70 s	10	0	104 s
VGALCD	N/A	N/A	N/A	N/A	N/A	N/A	4	0	985 s

In order to compare the performance of the BMC-based technique with our simulation-based methods, we applied the former directly, to minimize the original bug traces generated by semi-formal verification and by constrained-random simulation. For this experiment, the time-out limit was set to 40,000 seconds. Results are summarized in Table 8.10, where benchmarks that timed-out are marked by "T/O". The findings reported in the table confirm that our BMC-based method should only be applied, if at all, after the simulation-based techniques have already greatly reduced the trace complexity.

8.5.6 Evaluation of Experimental Results

We attempted to gain more insights into the experimental results by evaluating two additional aspects of the minimized traces. We first checked how close the minimized traces are to optimal-length traces such as those generated by formal verification. To do so, we run full-fledged SAT-based BMC on our minimized traces. The results show that our techniques found minimal-length bug traces for all benchmarks except DES (both traces generated by random simulation and semi-formal verification). For those two traces, the SAT solver ran out of memory after we unrolled the design by 118 cycles, and we could not finish the experiment. No shorter traces were found between 1 and 118 cycles long.

We also tried to evaluate if the potential for simulation-based trace reduction was mostly due to a large number of bug states, that is, a high number of design configurations that expose a given bug (an example of this situation is provided in Figure 8.1). To evaluate this aspect, we considered the original non-minimized traces in our experimental results. We first sampled the final

Table 8.10. Analysis of a pure BMC-based minimization technique.

Circuit	Original	Remained	Runtime (s)
S38584	13	9	403
S15850	59	59	338
MULT	345	T/O	T/O
DES	198	T/O	T/O
B15	25015	T/O	T/O
FPU	53711	T/O	T/O
ICU	6994	700	856
picoJava	30016	T/O	T/O
FPU	1046188	T/O	T/O
picoJava	99026	T/O	T/O
VGALCD	36595	T/O	T/O

This table shows the potential for minimizing traces using our BMC-based technique alone. Column "Original" shows the length, in cycles of the original trace, and column "Remained" shows the length of the minimized trace obtained after applying the BMC-based method. Traces in the top-half were generated by semi-formal verification, and the ones in the bottom-half were generated by constrained-random simulation. Experiments are timed-out at 40,000 seconds. The results of this table should be compared with Tables 8.3 and 8.5.

state of the design after simulating the traces, and then we fixed the goal of Butramin to generate a minimized trace that reaches that exact same final state. The results of this experiment are summarized in Table 8.11. The table shows that, for most benchmarks, the difference in the number of input events and cycles removed is small, showing that the size of the bug configuration has a minimal impact on the ability of Butramin to reduce and simplify a given bug trace, and our proposed solution remains effective even when the bug configuration is very specific.

8.6 Summary

In this chapter we presented Butramin, a bug trace minimizer that combines simulation-based techniques with formal methods. Butramin applies simple but powerful simulation-based bug trace reductions, such as *cycle elimination*, *input-event elimination*, *alternative path to bug*, *state skip* and *essential variable identification*. An additional BMC-based refinement method is used after these techniques to exploit the potential for further minimization. Compared to purely formal methods, Butramin has the following advantages: (1) it can reduce both the length of a bug trace and the number of its input events; (2) it leverages fast logic-simulation engines for bug trace minimization and it can scale to industrial size designs; and (3) it leverages existing simulation-based infrastructure, which is currently prevalent in the industry. This significantly

Table 8.11. Analysis of the impact of a bug radius on Butramin effectiveness.

Circuit	Cycles			Input events		
	Original trace	Same bug	Same state	Original trace	Same bug	Same state
S38584	13	8	9	255	2	41
S15850	59	1	1	2300	3	3
MULT	345	4	4	43843	2	380
DES	198	154	193	3293	3	1022
B15	25015	11	11	450026	15	40
FPU	53711	5	5	1756431	17	112
ICU	6994	3	5	62740	3	6
picoJava	30016	10	75	675485	11	1575
FPU	1046188	5	6	36125365	17	120
picoJava	99026	10	22	2227599	16	42
VGALCD	36595	4	199	1554616	19	2068

The table compares number of cycles and input events in the original traces to the same values from minimized traces that hit the same bug, and to minimized traces that reach the same bug configuration. Traces in the top-half were generated by semi-formal software and traces in the bottom-half were generated by constrained-random simulation.

lowers the barriers for industrial adoption of automatic design verification techniques.

Our experimental results show that Butramin can reduce a bug trace to just a small fraction of its original length and complexity (estimated as the number of input events in the trace) by using only simulation-based techniques. In addition we showed that these results are largely independent of the verification methodology used to generate the trace, whether based on simulation or semi-formal verification techniques. The impact of Butramin appears to be uncorrelated with the size of the bug configuration targeted by the trace, that is, the number of distinct design states that expose the bug.

Recent follow-up work by Pan et al. [53] and Safarpour et al. [115] focuses on improving the formal analysis techniques for bug trace minimization, and their approaches can be used to augment our BMC-based technique. As their experimental results suggest, however, formal analysis still cannot achieve the scalability provided by our simulation-based minimization methods, making Butramin more suitable for practical designs.

Chapter 9

FUNCTIONAL ERROR DIAGNOSIS AND CORRECTION

Recent improvements in design verification strive to automate the error-detection process and greatly enhance engineers' ability in detecting the presence of functional errors. However, the process of diagnosing the cause of these errors and fixing them remains difficult and requires significant manual effort. The work described in this chapter improves this aspect of verification by presenting new constructs and algorithms to automate the error-repair process at both the gate level and the Register-Transfer Level (RTL). In this chapter, we first extend the CoRé framework (see Chapter 5) to handle sequential circuits. Next, we present a novel RTL error diagnosis and correction methodology. Finally, we show the empirical evaluation of our functional error repair techniques and summarize this chapter.

9.1 Gate-Level Error Repair for Sequential Circuits

The CoRé framework described in Chapter 5 only addresses the error-repair problem for combinational circuits. CoRé is easily adaptable to correct errors in sequential circuits, as described in this section. First of all, when operating on sequential circuits the user will provide CoRé with input traces, instead of input patterns. A trace is a sequence of input patterns, where a new pattern is applied to the design's inputs at each simulation cycle, and the trace can be either *error-sensitizing* or *functionality-preserving*. To address sequential circuits, we adopt the diagnosis techniques from Ali et al. [6] relating to sequential circuits. The idea is to first unroll the circuit by connecting the outputs of the state registers to the inputs of the registers in the previous cycle, and then use the test vectors to constrain the unrolled circuit. Given an initial state and a set of test vectors with corresponding correct output responses, Ali's error-diagnosis technique is able to produce a collection of error sites, along with their correct values, that rectify the incorrect output responses.

To correct errors in sequential designs we apply the same algorithm described in Section 5.2.1 with two changes: the diagnosis procedure should be as described in [6], and the signature generation function is modified so that it can be used in a sequential design. Specifically, the new sequential signature generation procedure should record one bit of signature for each cycle of each sequential trace that we simulate. For instance, if we have two traces available, a 4-cycle trace and a 3-cycle trace, we will obtain a 7-bit signature at each internal circuit node. An example of the modified signature is shown in Figure 9.1. In our current implementation, we only use combinational Observability Don't-Cares (ODCs). In other words, we still treat inputs of state registers as primary outputs when calculating ODCs. Although it is possible to exploit sequential ODCs for resynthesis, we do not pursue this optimization, yet.

	Trace1				Trace2		
Cycle	1	2	3	4	1	2	3
Signature	0	1	1	0	1	0	1

Figure 9.1. Sequential signature construction example. The signature of a node is built by concatenating the simulated values of each cycle for all the bug traces. In this example, trace1 is 4 cycles and trace2 is 3 cycles long. The final signature is then 0110101.

9.2 Register-Transfer-Level Error Repair

To develop a scalable and powerful RTL error diagnosis and correction system, we extend our gate-level techniques to the RTL. This approach is more accurate than previous software-based RTL solutions [75, 111, 122] (see Section 2.1) in that we can analyze designs rigorously using formal hardware verification techniques. At the same time, it is considerably faster and more scalable than gate-level diagnosis because errors are modeled at a higher level. Moreover, it only requires test vectors and output responses, making it more practical than existing formal analysis solutions [18]. Finally, the novel error model and increased accuracy of our approach allow our technique to provide insightful suggestions for correcting diagnosed errors. Key ideas in this work include: (1) a new RTL error model that explicitly inserts MUXes into RTL code for error diagnosis, as opposed to previous solutions that use MUXes implicitly; (2) new error-diagnosis algorithms using synthesis or symbolic simulation; and (3) an error-correction technique using signal behaviors (*signatures*) that are especially suitable for the RTL. Empirical results show that these techniques allow us to provide highly accurate diagnoses very quickly.

We implemented our techniques in a framework called *REDIR* (RTL Error DIagnosis and Repair), highlighted in Figure 9.2. The inputs to the framework include a design containing one or more bugs, a set of test vectors exposing them, and the correct responses for the primary outputs over the given test vectors (usually generated by a high-level behavioral model written in C, C++, SystemC, etc.). Note that we only require the correct responses at the primary

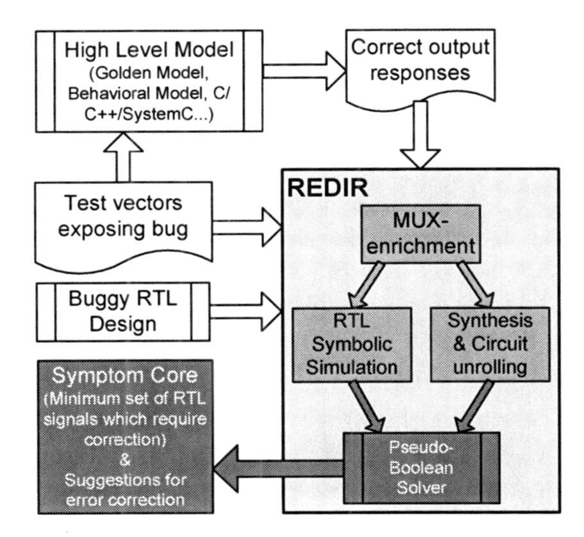

Figure 9.2. REDIR framework. Inputs to the tool are an RTL design (which includes one or more errors), test vectors exposing the bug(s), and correct output responses for those vectors obtained from high-level simulation. Outputs of the tool include REDIR *symptom core* (a minimum cardinality set of RTL signals which need to be modified in order to correct the design), as well as suggestions to fix the errors.

outputs of the high-level model and no internal values are required. The output of the framework is a minimum cardinality set of RTL signals that should be corrected in order to eliminate the erroneous behavior. We call this set the *symptom core*. When multiple cores exist, REDIR provides all of the possible minimal cardinality sets. In addition, the framework suggests several possible fixes of the signals in the *symptom core* to help a designer correct those signals.

The rest of the section is organized as follows. In Section 9.2.1, we provide the necessary background. Section 9.2.2 describes our error-diagnosis techniques, and Section 9.2.3 explains our error-correction method.

9.2.1 Background

Our error-diagnosis algorithm converts the error-diagnosis problem into a *Pseudo-Boolean (PB)* problem, and then uses a PB solver to perform the diagnosis and infer which design signals are responsible for incorrect output behavior. In this subsection, we first define Pseudo-Boolean problems, which are an extension of SATisifiability problems. Next, we review the basic idea behind symbolic simulation, which we use as an alternative, compact way to formulate the PB problem.

Pseudo-Boolean Problems

PB problems, also called 0–1 integer linear programming problems, are an extension of SATisfiability problems. A Pseudo-Boolean Constraint (PBC) is specified as an inequality with a linear combination of Boolean variables: $C_0p_o + C_1p_1 + ... + C_{n-1}p_{n-1} \geq C_n$, where the variables p_i are defined over the Boolean set $\{0, 1\}$. A PB problem allows the use of an additional *objective function*, which is a linear expression that should be minimized or maximized under the given constraints. A number of PB solvers have been developed recently by extending existing SAT solvers (for instance, MiniSat+ [62]).

Logic and Symbolic Simulation

Logic simulation models the behavior of a digital circuit by propagating scalar Boolean values (0 and 1) from primary inputs to primary outputs. For example, when simulating 2-input AND with both inputs set to 1, the output 1 is produced. On the other hand, *symbolic simulation* uses symbols instead of scalar values and produces Boolean expressions at the outputs [15, 22]. As a result, simulating a 2-input XOR with inputs a and b generates an expression "a XOR b" instead of a scalar value. To improve scalability, modern symbolic simulators employ several techniques, including approximation, parameterization and on-the-fly logic simplification [14]. For example, with on-the-fly logic simplification, "*0* XOR b" is simplified to b thus reducing the complexity of the expression. Traditional symbolic simulators operate on a gate-level model of a design; however, in recent years simulators operating on RTL descriptions have been proposed [82, 83]. Symbolic simulation is an alternative way to generate an instance of the PB constraint problem that we use in our error-diagnosis framework.

9.2.2 RTL Error Diagnosis

In this subsection, we describe our error-diagnosis techniques. First, we explain our RTL error model, and then propose two diagnosis methods that use either synthesis or symbolic simulation. Finally, we outline how hierarchical designs should be handled.

Error Modeling

In our framework the error-diagnosis problem is represented with (1) an RTL description containing one or more bugs that is composed of variables (wires, registers, inputs/outputs) and operations on those variables; (2) a set of test vectors exposing the bugs; and (3) the correct output responses for the given test vectors, usually generated by a high-level behavioral model. The objective of the error diagnosis is to identify a minimal number of variables in the RTL description that are responsible for the design's erroneous behavior.

Moreover, by modifying the logic of those variables, the design errors can be corrected. Each signal found to affect the correctness of the design is called a *symptom variable*. Without minimization, the set of symptom variables reported would include the root cause of the bug and the cone of logic emanating from it: correcting all the symptom variables on any cut across this cone of logic would eliminate the bug. Therefore, by forcing the PB solver to minimize the number of symptom variables, we return a solution as close to the root cause of the erroneous behavior as possible.

To model errors in a design, we introduce a conditional assignment for each RTL variable, as shown in the example in Figure 9.3. Note that these conditional assignments are used for error diagnosis only and should not appear in the final synthesized design. However, they allow the REDIR framework to locate sites of erroneous behavior in RTL, as we illustrate using a *half_adder* design shown in Figure 9.3. Suppose that the output responses of the design are incorrect because c should be driven by "a & b" instead of "a | b". Obviously, to produce the correct output that we obtain from a high-level model, the behavior of c must be changed. To model this situation, we insert a conditional assignment, "assign $c_n = c_{sel}$? c_f : c", into the code. Next, we replace all occurrences of c in the code with c_n, except when c is used on the left-hand-side of an assignment. We call c_{sel} a *select variable* and c_f a *free variable*. Then, by asserting c_{sel} and using an alternative signal source, modeled by c_f, we can force the circuit to behave as desired. If we can identify the select variables that should be asserted and the correct signals that should drive the corresponding free variables to produce correct circuit behavior, we can diagnose and fix the errors in the design.

The procedure to introduce a conditional assignment for a design variable v is called MUX-enrichment (since conditional assignments are conceptually multiplexers), and its pseudo-code is shown in Figure 9.4. It should be performed on each internal signal, defined in the circuit, including registers. The primary inputs, however, should not be MUX-enriched since by construction they cannot have erroneous values. It also should be noted that for hierarchical designs the primary inputs of a module may be driven by the outputs of another module and, therefore, may assume erroneous values. To handle this situation, we insert conditional assignments into the hierarchical modules' output ports.

Diagnosis with Synthesis

After the error-modeling constructs have been inserted into a design, error diagnosis is used to identify the minimal number of select variables that should be asserted along with the values of their corresponding free variables to produce the correct circuit behavior. In this section we present an error-diagnosis technique that uses synthesis and circuit unrolling. In contrast with existing gate-level diagnosis techniques described in Section 5.1.3, our RTL error-modeling constructs are synthesized with the design, which eliminates the need to insert multiplexers at the gate level. In this way, the synthesized

```
module half_adder(a, b, s, c);
  input a, b;
  output s, c;
  assign s = a ^ b;
  assign c = a | b;
endmodule

module half_adder_MUX_enriched(a, b, s_n, c_n, s_sel, c_sel, s_f, c_f);
  input a, b, s_sel, c_sel, s_f, c_f;
  output s_n, c_n;
  assign s = a ^ b;
  assign c = a | b;
  assign s_n = s_sel ? s_f : s;
  assign c_n = c_sel ? c_f : c;
endmodule
```

Figure 9.3. An RTL error-modeling code example: module half_adder shows the original code, where c is erroneously driven by "$a \mid b$" instead of "a & b"; and module half_adder_MUX_enriched shows the MUX-enriched version. The differences are marked in boldface.

procedure $MUX\_enrichment(v)$
1. create a new signal wire v_n and new inputs v_f and v_{sel};
2. add conditional assignment "$v_n = v_{sel}\ ?\ v_f : v$";
3. replace all occurrences of v that appear on the right-hand-side of assignments (including outputs, if/case conditions, etc.) with v_n;

Figure 9.4. Procedure to insert a conditional assignment for a signal in an RTL description for error modeling.

netlist faithfully preserves the constructs inserted at the RTL, enabling accurate RTL error diagnosis. This is significantly different from diagnosing design errors at the gate level, since synthesis is only used to generate Boolean expressions between RTL variables, and the synthesized netlist is not the target of the diagnosis. As a result, our diagnosis method has a much smaller search space

Procedure $syn\_based\_diagnosis(designCNF, c, inputs, outputs)$
1. CNF = unroll $designCNF$ c times;
2. connect all *select variables* in CNF to those in the first cycle;
3. constrain PI/PO in CNF using $inputs/outputs$;
4. $PBC = CNF$, min($\sum$ *select variables*);
5. return solution= PB-Solve(BPC);

Figure 9.5. Procedure to perform error diagnosis using synthesis and circuit unrolling. PI/PO means primary inputs and primary outputs.

and runs significantly faster than gate-level techniques, as we show in the experimental results.

Figure 9.5 outlines the algorithm for synthesis-based error diagnosis. Before the procedure is called, the design is synthesized and its combinational portion is converted to CNF format ($designCNF$). Other inputs to the procedure include the length of the bug trace, c, as well as the test vectors ($inputs$) and their correct output responses ($outputs$). To make sure that the diagnosis applies to all simulation cycles, the algorithm connects the select variables for each unrolled copy to the corresponding CNF variables in the first copy. On the other hand, free variables for each unrolled copy of the circuit are independent. When a solution is found, each asserted select variable is a symptom variable, and the solution for its corresponding free variable is an alternative signal source that can fix the design errors. Note that if state values over time are known, they can be used to constrain the CNF at register boundaries, reducing the sequential error-diagnosis problem to combinational. The constructed CNF, along with the objective to minimize the sum of select variables, forms a PBC. Error diagnosis is then performed by solving the PBC.

Diagnosis with RTL Symbolic Simulation

Here we propose an alternative error-diagnosis technique that scales further than the synthesis-based technique. We achieve this by performing symbolic simulation directly on the RTL representation and generating Boolean expressions at the primary outputs for all simulated cycles. The outputs' Boolean expressions are used to build a PB problem's instance, that is then handed over to a PB solver for error diagnosis.

Although RTL symbolic simulators are not yet commonly available in the industry, effective solutions have been proposed recently in the literature [82, 83]. Moreover, because of the scalability advantages of performing symbolic simulation at the RTL instead of the gate level, commercial-quality solutions are starting to appear. For our empirical validation we used one such RTL symbolic simulator [147].

Figure 9.6 illustrates our procedure that uses symbolic simulation and PB solving. We assume that the registers are initialized to known values before the procedure is invoked. We also assume that the circuit contains n MUX-enriched signals named v_i, where $i = \{1..n\}$. Each v_i has a corresponding select variable v_{i_sel} and a free variable v_{i_f}. There are o primary outputs, named PO_j, where $j = \{1..o\}$. We use subscript "@" to prefix the cycle during which the symbols are generated. For each primary output j and for each cycle t we compute expression $PO_{j@t}$ by symbolically simulating the given RTL design, and also obtain correct output value $CPO_{j@t}$ from the high-level model. The inputs to the procedure are the RTL design ($design$), the test vectors ($test\_vectors$), and the correct output responses over time (CPO).

```
Procedure sim_based_diagnosis(design, test_vectors, CPO)
1  ∀i, 1 ≤ i ≤ n, v_i_sel = new_symbol();
2  for t = 1 to c begin // Simulate c cycles
3     PI = test_vector at cycle t;
4     ∀i, 1 ≤ i ≤ n, v_i_f@t = new_symbol();
5     PO@t = simulate(design);
6  end
7  PBC = ∧_{j=1}^{o} ∧_{t=1}^{c} (PO_j@t == CPO_j@t), min(∑_{i=1}^{n} v_i_sel);
8  return solution= PB_Solve(PBC);
```

Figure 9.6. Procedure to perform error diagnosis using symbolic simulation. The boldfaced variables are symbolic variables or expressions, while all others are scalar values.

In the algorithm shown in Figure 9.6, a symbol is initially created for each select variable (line 1). During the simulation, a new symbol is created for each free variable in every cycle, and test vectors are applied to primary inputs, as shown in lines 2–4. The reason for creating only one symbol for each select variable is that a conditional assignment should be either activated or inactivated throughout the entire simulation, while each free variable requires a new symbol at every cycle because the value of the variable may change. As a result, the symbols for the select variables are assigned outside the simulation loop, while the symbols for the free variables are assigned in the loop. The values of the free variables can be used as the alternative signal source to produce the correct behavior of the circuit. After simulating one cycle, a Boolean expression for all of the primary outputs are created and saved in $PO_{@t}$ (line 5). After the simulation completes, the generated Boolean expressions for all the primary outputs are constrained by their respective correct output values and are ANDed to form a PBC problem as line 7 shows. In order to minimize the number of symptom variables, we minimize the *sum* of select variables, which is also added to the PBC as the objective function. A PB solver is then invoked to solve the formulated PBC, as shown in line 8. In the solution, the asserted select variables represent the symptom variables, and the values of the free variables represent the alternative signal sources that can be used to correct the erroneous output responses.

Below we present an example of a buggy design to illustrate the symbolic simulation-based error-diagnosis technique.

EXAMPLE 9.1 *Assume that the circuit shown in Figure 9.7 contains an error: signal g_1 is erroneously assigned to expression "r1 | r2" instead of "r1 & r2". Conditional assignments, highlighted in boldface, have been inserted into the circuit using the techniques described earlier. For simplicity reasons, we do not include the MUXes at the outputs of registers r1 and r2. The trace that exposes the error in two simulation cycles consists of the following val-*

```
module example(clk, I1, I2, O1_n, O2_n, g1_sel, O1_sel, O2_sel, g1_f, O1_f, O2_f);
  input I1, I2, g1_sel, O1_sel, O2_sel, g1_f, O1_f, O2_f
  output O1_n, O2_n;
  reg r1, r2;
  initial begin r1= 0; r2= 0; end
  always @(posedge clk) begin
    r1= I1; r2= I2;
  end
  assign g1 = r1 | r2;
  assign O1 = I1 | g1_n;
  assign O2 = I2 & g1_n;
  assign g1_n= g1_sel ? g1_f : g1;
  assign O1_n= O1_sel ? O1_f : O1;
  assign O2_n= O2_sel ? O2_f : O2;
endmodule
```

Figure 9.7. Design for the example. Wire g1 should be driven by "r1 & r2", but it is erroneously driven by "r1 | r2". The changes made during MUX-enrichment are marked in boldface.

ues for inputs {I1, I2}: {0, 1}, {1, 1}. When the same trace is simulated by a high-level behavioral model, the correct output responses for {O1, O2} are generated: {0, 0}, {1, 0}. Besides these output responses, no addition information, such as values of internal signals and registers, is required. We annotate the symbols injected during the simulation by their cycle numbers using subscripts. The Boolean expressions for the primary outputs for the two cycles of simulation are:

$O1_{n@1} = O1_{sel} \ ?\ O1_{f@1} : [I1_{@1} \mid (g1_{sel} \ ?\ g1_{f@1} : 0)]$

$O2_{n@1} = O2_{sel} \ ?\ O2_{f@1} : [I2_{@1} \ \&\ (g1_{sel} \ ?\ g1_{f@1} : 0)]$

$O1_{n@2} = O1_{sel} \ ?\ O1_{f@2} : \{I1_{@2} \mid [g1_{sel} \ ?\ g1_{f@2} : (I1_{@1} \ \&\ I2_{@1})]\}$

$O2_{n@2} = O2_{sel} \ ?\ O2_{f@2} : \{I2_{@2} \ \&\ [g1_{sel} \ ?\ g1_{f@2} : (I1_{@1} \ \&\ I2_{@1})]\}$

Since the primary inputs are scalar values, the expressions can be greatly simplified during symbolic simulation. For example, we know that $I1_{@2}=1$; therefore, $O1_{n@2}$ can be simplified to $O1_{sel} \ ?\ O1_{f@2} : 1$. As a result, the Boolean expressions actually generated by the symbolic simulator are:

$O1_{n@1} = O1_{sel} \ ?\ O1_{f@1} : (g1_{sel} \ ?\ g1_{f@1} : 0)$

$O2_{n@1} = O2_{sel} \ ?\ O2_{f@1} : (g1_{sel} \ ?\ g1_{f@1} : 0)$

$O1_{n@2} = O1_{sel} \ ?\ O1_{f@2} : 1$

$O2_{n@2} = O2_{sel} \ ?\ O2_{f@2} : (g1_{sel} \ ?\ g1_{f@2} : 0)$

To perform error diagnosis, we constrain the output expressions using the correct responses, and then construct a PBC as follows:

$PBC = (O1_{n@1} == 0) \wedge (O2_{n@1} == 0) \wedge (O1_{n@2} == 1) \wedge (O2_{n@2} == 0)$,
$min(O1_{sel} + O2_{sel} + g1_{sel})$.

One solution of this PBC is to assert $g1_{sel}$, which provides a correct symptom core.

Handling Hierarchical Designs

Current designs often have hierarchical structures to allow the circuits to be decomposed into smaller blocks and thus reduce their complexity. Here we discuss how a MUX-enriched circuit should be instantiated if it is encapsulated as a module in such a hierarchical design.

The algorithm to insert MUXes into a single module m is shown in Figure 9.4. If m is instantiated inside of another module M, however, MUX-enrichment of M must include an extra step where new inputs are added to all instantiations of m. Therefore, for hierarchical designs, the insertion of conditional assignments must be performed bottom-up: MUX-enrichment in a module must be executed before it is instantiated by another module. This is achieved by analyzing the design hierarchy and performing MUX-enrichment in a reverse-topological order.

It is important to note that in hierarchical designs, the select variables of instances of the same module should be shared, while the free variables should not. This is because all instances of the same module will have the same symptom variables. As a result, select variables should share the same signals. On the other hand, each instance is allowed to have different values for their internal signals; therefore, each free variable should have its own signal. However, it is possible that a bug requires fixing only one RTL instance while other instances of the same module can be left intact. This situation requires generation of new RTL modules and is currently not handled by our diagnosis techniques.

9.2.3 RTL Error Correction

The RTL error-correction problem is formulated as follows: given an erroneous RTL description of a digital design, find a variant description for one or more of the modules that compose it so that the new design presents a correct behavior for the errors, while leaving the known-correct behavior unchanged. Although many error-repair techniques exist for gate-level designs, very few studies focus on the RTL. One major reason is the lack of logic representations that can support the logic manipulation required during RTL error correction. For example, the logic of a signal in a gate-level netlist can be easily represented by BDDs, and modifying the function of the signal can be supported by the manipulation of its BDDs. However, most existing logic representations cannot be easily applied to an RTL variable. This problem is further exacerbated by the fact that an RTL module may be instantiated multiple times, creating many different functions for an RTL variable depending on where it is instantiated.

In this subsection, we first describe the baseline error-correction technique that is easier to understand. Next, we show how signatures should be generated at the RTL to handle hierarchical and sequential designs. Finally, we provide some insights that we obtained during the implementation of our system.

Baseline Error-Correction Technique

For a flattened combinational design, error correction is performed as follows: (1) signatures of RTL variables are generated using simulation; (2) error diagnosis is performed to find a symptom core; (3) signatures of the symptom variables in the symptom core are replaced by the values of their corresponding free variables; and (4) synthesis is applied to find logic expressions generating the signatures of the symptom variables. By replacing the expressions that generate the functions of the symptom variables with those new expressions, design errors can be corrected.

Hierarchical and Sequential Designs

In a flattened design, each RTL variable represents exactly one logic function. In a hierarchical design, however, each variable may represent more than one logic function. Therefore, we devise the following techniques to construct the signatures of RTL variables. For clarity, we call a variable in an RTL module a *module variable* and a variable in an instance generated by the module an *instance variable*. A module variable may generate multiple instance variables if the module is instantiated several times.

In RTL error correction, we modify the source code of the modules in order to correct the design's behavior. Since changing an RTL module will affect all the instances produced by the module, we concatenate the simulation values of the instance variables derived from the same module variable to produce the signature for the module variable. This way, we can guarantee that a change in a module will affect instances in the same way. Similarly, we concatenate the signatures of the module variable at different cycles for sequential error correction. A signature-construction example is given in Figure 9.8. Note that to ensure the correctness of error repair, the same instance and cycle orders must be used during the concatenation of signatures for all module variables.

EXAMPLE 9.2 *Using the same circuit as Example 9.1. The values returned by the PB solver for* $g1_{f@0}$ *and* $g1_{f@1}$ *are both 0. Since the inputs to* $g1$ *are* $\{0, 0\}$ *and* $\{0, 1\}$ *for the first two cycles, the correct expression for* $g1$ *should generate 0 for these two inputs. RTL error correction returns the following new logic expressions that can fix the error:* $g1 = r1\&r2$, $g1 = r1$, *etc. Note that although the correct fix is returned, the fix is not unique. In general, longer traces containing various test vectors will identify the error with higher precision and suggest better fixes than short ones.*

Design:
```
module top;
  child c1(), c2(), c3();
endmodule
module child;
  wire v;
endmodule
```
Simulation values:

Cycle 0: top.c1.v = 0, top.c2.v = 0, top.c3.v = 1

Cycle 1: top.c1.v = 1, top.c2.v = 0, top.c3.v = 0

Constructed signature for RTL error correction:

$$\text{child.v} = \underbrace{\overbrace{1}^{c1.v}\ \overbrace{0}^{c2.v}\ \overbrace{0}^{c3.v}}_{cycle\ 1}\ \underbrace{\overbrace{0}^{c1.v}\ \overbrace{0}^{c2.v}\ \overbrace{1}^{c3.v}}_{cycle\ 0}$$

Figure 9.8. Signature-construction example. Simulation values of variables created from the same RTL variable at all cycles should be concatenated for error correction.

Identifying Erroneous Code Statements

Several existing error-diagnosis techniques are able to identify the RTL code statements that may be responsible for the design errors [76, 111, 122, 127]. Unlike these techniques, REDIR returns the RTL variables that are responsible for the errors instead. Since one variable may be affected by multiple statements, the search space of the errors modeled by these techniques tend to be larger than REDIR, making REDIR more efficient in error diagnosis. On the other hand, being able to identify erroneous statements may further localize the errors and make debugging easier. To achieve this goal, we observe that in correctly designed RTL code, the value of a variable should be affected by at most one statement at each cycle. Otherwise, a multiple-driven error will exist in the design. Based on this observation, we develop the following procedure to identify the erroneous code statements using our error-diagnosis results.

Given a symptom variable, we first record the cycles at which the values of its free variables are different from its simulated values. Next, we identify the code statements that assign new values to the symptom variable for those cycles: these code statements are responsible for the errors. Since many modern logic simulators provide the capability to identify the executed code statements (e.g., full-trace mode in Cadence Verilog-XL), erroneous statements can be pinpointed easily by replaying the bug traces used for error diagnosis. After erroneous statements are identified, signatures for error-correction can be generated using only the cycles when the statements are executed. In this way, we can produce corrections specifically for the erroneous statements.

Implementation Insights

When multiple bug traces are used in the diagnosis, the set of the reported symptom variables is the intersection of the symptoms identified by each bug trace. Therefore, to accelerate the diagnosis over a specific bug trace, we can deassert the select variables that are never asserted during the execution of previous traces.

Fixing errors involving multi-bit variables is more difficult than fixing errors involving only one-bit variables because different bits in the variable may be generated differently. To solve this problem, we allow the user to insert a conditional assignment for each bit in the variable. Alternatively, REDIR can also be configured to consider only the least-significant bit when performing error correction. This is useful when the variable is considered as a whole.

In synthesis-based error diagnosis, we observe that it is difficult to identify the wires derived from the same RTL variable in a synthesized netlist. To overcome this problem, we add the outputs of inserted conditional statements to the primary outputs of the MUX-enriched modules to obtain the simulated values of the RTL variables. To improve our error-correction quality, we utilize ODCs in our synthesis-based approach by simulating the complement signatures of symptom variables and observe the changes at primary outputs (including inputs to registers).

9.3 Experimental Results

In this section we present experimental results of our error-repair techniques at both the gate level and the RTL. At the gate level, we first evaluate the effectiveness of the baseline CoRé framework on fixing bugs in combinational designs. Next, we use the extension described in Section 9.1 to repair errors in sequential designs. At the RTL, we first evaluate our error-diagnosis techniques and contrast the results with those at the gate level. We then show the results on automatic error correction.

9.3.1 Gate-Level Error Repair

We implemented our CoRé framework using the OAGear package [162] because it provides convenient logic representations for circuits. We adopted Smith's [125] algorithm and integrated MiniSat [61] into our system for error diagnosis and equivalence checking. We used Espresso [114] to optimize the truth table returned by DPS, and then we constructed the resynthesized netlist using AND, OR and INVERTER gates. Our testcases were selected from IWLS2005 benchmarks [161] based on designs from ISCAS89 and OpenCores suites. In our implementation, we limited the number of attempts to resynthesize a wire to 30, and we prioritized our correction by starting from fixes with wires closer to primary inputs. We conducted four experiments on a 2.0GHz

Pentium 4 workstation. The first two experiments are in the context of equivalence checking, the third one deals with simulation-based verification, while the last one repairs errors in sequential circuits.

Equivalence checking: our first experiment employs Application 1 described in Section 5.2.4 to repair an erroneous netlist by enforcing equivalency. Inputs and outputs of the sequential elements in the benchmarks were treated as primary outputs and inputs, respectively. The initial vectors were obtained by simulating 1024 random patterns, and one error was injected to each netlist. In the first half of the experiment, the injected errors fit in the error model described in Section 5.1.4; while the errors injected in the second half involved more than 2 levels of logic and did not comply with the error model. We applied GDS and DPS separately to compare their error-correction power and performance. Since GDS subsumes existing techniques that are based on error models, it can be used as a comparison to them. The results are summarized in Table 9.1. As expected, GDS could not repair netlists in the second half of the experiment, showing that our resynthesis techniques could fix more errors than those based on Abadir's error models [86, 130].

From the results in the first half, we observe that both GDS and DPS performed well in the experiment: the resynthesis time was short, and the number of iterations was typically small. This result shows that the error-diagnosis technique we adopted was effective and our resynthesis techniques repaired the netlists correctly. Compared with the error-correction time required by some previous techniques that enumerate possible fixes in the error model [54, 130], the short runtime of GDS shows that our pruning methods are efficient, even though GDS also explores all possible combinations. We observe that the program runtime was dominated by error diagnosis and verification, which highlights the importance of developing faster error-diagnosis and verification techniques.

Errors that are difficult to diagnose and correct often need additional test vectors and iterations. In order to evaluate our techniques on fixing difficult errors, we reran the first three benchmarks and reduced the number of their initial patterns to 64. The results are summarized in Table 9.2, where the number of iterations increased as expected. The results suggest that our techniques continued to be effective for difficult errors, where all the errors could be fixed within two minutes. We also observe that DPS may sometimes need more iterations due to its much larger search space. However, our framework would guide both techniques to the correct fix eventually.

In our second experiment, we injected more than one error into the netlist. The injected errors complied with Abadir's model and could be fixed by both GDS and DPS. To mimic difficult errors, the number of initial vectors was reduced to 64. We first measured the runtime and the number of iterations required to fix each error separately, we then showed the results on fixing

Table 9.1. Error-correction experiment for combinational gate-level netlists.

Bench-mark	Gate count	Type of injected error	GDS				DPS			
			Runtime (sec)			No. of iter.	Runtime (sec)			No. of iter.
			EC	ED	Veri.		EC	ED	Veri.	
S1488	636	Single gate change	1	3	1	1	1	4	1	1
S15850	685	Connection change	1	5	1	2	2	5	1	1
S9234_1	974	Single gate change	1	10	1	1	1	9	1	1
S13207	1219	Connection change	1	5	1	1	1	5	1	1
S38584	6727	Single gate change	1	306	83	1	1	306	81	1
S838_1	367	Multiple gate changes	N/A				1	6	1	1
S13207	1219	Multiple missing gates	N/A				3	12	3	6
AC97_ctrl	11855	Multiple connection changes	N/A				2	1032	252	5

The benchmarks in the top-half comply with Abadir's error model, while those in the bottom-half do not. "No. of iter." is the number of error-correction attempts processed by the verification engine. "EC" means error correction, "ED" means error diagnosis, and "Veri." means verification.

Table 9.2. Error-correction experiment for combinational gate-level netlists with reduced number of initial patterns.

Bench-mark	Gate count	Type of injected error	GDS				DPS			
			Runtime (sec)			No. of iter.	Runtime (sec)			No. of iter.
			EC	ED	Verification		EC	ED	Verification	
S1488	636	Single gate change	1	5	3	13	1	4	1	3
S15850	685	Connection change	1	3	1	5	53	4	5	42
S9234_1	974	Single gate change	1	8	3	6	1	10	3	4

multiple errors. Time-out was set to 30 minutes in this experiment, and the results are summarized in Table 9.3. Similar to other error diagnosis and correction techniques, runtime of our techniques grows significantly with each additional error. However, we can observe from the results that the number of iterations is usually smaller than the product of the number of iterations for each error. It shows that our framework tends to guide the resynthesis process to fix the errors instead of merely trying all possible combinations of fixes. Another interesting phenomenon is that DPS could simultaneously fix all three errors in the S1488 benchmark, while GDS could not. The reason is that DPS found a fix involving only two wires even though three errors were injected. Since GDS could not fix the netlist using only two error sites, three-error diagnosis was performed, which was extremely slow. The reason is that in addition to fixes involving three error sites, any combination of wires consisting of two error sites and one "healthy" site (site with its function unchanged) is also a valid fix. As a result, the number of possible fixes increased dramatically and evaluating all of them was time consuming. This explanation is confirmed by the following observation: error diagnosis returned 8, 7 and 9 possible fixes for error1, error2 and error3 respectively, while the number of fixes for all three errors using three sites was 21,842. This situation suggests that DPS is more powerful than GDS, as well as many techniques subsumed by GDS.

Table 9.3. Multiple error experiment for combinational gate-level netlists.

Benchmark	Runtime (sec)					Number of iterations				
	Err1	Err2	Err3	Err1+2	Err1+2+3	Err1	Err2	Err3	Err1+2	Err1+2+3
S1488(GDS)	4	6	4	10	T/O	8	5	2	22	T/O
S1488(DPS)	14	5	5	34	9	32	4	2	45	14
S13207(GDS)	10	10	6	12	75	11	5	1	10	19
S13207(DPS)	7	9	6	14	74	4	5	1	16	15
S15850(GDS)	4	3	4	5	7	1	1	1	1	1
S15850(DPS)	4	3	5	5	10	1	1	13	1	11

Time-out is set to 30 minutes and is marked as T/O in the Table.

To further evaluate the strength of our error-repair techniques, we took the C17 benchmark from the ISCAS'85 suite and prepared a totally different circuit with the same number of primary inputs and outputs, where the circuit is composed of two multiplexers. Next, we used CoRé to "repair" the C17 benchmark so that it became equivalent to the prepared circuit. CoRé successfully repaired the C17 benchmark in 0.04 seconds using 26 test vectors. Since the number of primary inputs is 5, this result suggests that CoRé can find a fix without trying all possible input patterns (32 in this benchmark), even when the the repaired circuit is considerably different from the original one.

Simulation-based verification: in our third experiment, we simulated n functionality-preserving vectors and m error-sensitizing vectors, where m is much smaller than n. Error-sensitizing vectors were produced by randomly changing one output per vector. We then checked whether our framework could produce a netlist that was adaptive to the new responses. This is similar to fixing errors found by simulation-based verification, where a few vectors break the regression test while most vectors should be preserved. In this experiment, we set n=1024 while changing m, and the results are summarized in Table 9.4. We can observe from the results that additional error-sensitizing vectors usually require more wires to be fixed, and the runtime is also longer. However, our framework is able to repair all the benchmarks within a short time by resynthesizing only a small number of wires. This result suggests that our framework works effectively in the context of simulation-based verification.

Table 9.4. Error correction for combinational gate-level netlists in the context of simulation-based verification.

Bench-	Runtime (sec)				Number of error sites			
mark	m=1	m=2	m=3	m=4	m=1	m=2	m=3	m=4
S1488	3	4	10	10	1	2	3	3
S15850	3	4	4	6	1	2	2	4
S13207	3	6	8	19	1	2	3	5

1024 functionality-preserving and m error-sensitizing vectors are simulated, where the error-sensitizing vectors randomly change one output per vector.

Repairing errors in sequential circuits: our fourth experiment repairs errors in sequential circuits using techniques described in Section 9.1. The characteristics of the benchmarks and their results are summarized in Table 9.5. For each benchmark, 32 traces were provided, and the goal was to repair the circuit so that it produces the correct output responses for those traces. Note that diagnosing errors in sequential circuits is much more difficult than that in combinational circuits because circuit unrolling is used. For example, the bug trace for the last benchmark had 77 cycles, and it produced an unrolled circuit containing more than one million standard cells. Since our algorithm processes all the traces simultaneously, only one iteration will be required. For the computation of more representative runtime only, we deliberately processed the traces one by one and failed all verification so that all the benchmarks underwent 32 iterations. All the bugs were injected at the RTL, and the designs were synthesized using Cadence RTL compiler 4.10. In the table, "Err. Diag. time" is the time spent on error diagnosis, "#Fixes" is the number of valid fixes returned by CoRé, and "DPS time" is the runtime of DPS. The minimum/maximum numbers of support variables and gates used in the returned fixes are shown under "Resynthesized netlist". Note that implementing any valid fix is sufficient to correct the circuit's behavior, and we rank the fixes based on the logic depth from primary inputs: fixes closer to primary inputs are preferred. Under "Err. diag. time", "1st" is the runtime for diagnosing the first bug trace, while "Total" is the runtime for diagnosing all 32 traces.

Table 9.5. Error-repair results for sequential circuits.

Bench-mark	Descrip-tion	#Cells	Bug description	Err. diag. time (sec)		#Fix-es	Resynthesized netlist		DPS time (sec)
				First	Total		#Supp.	#Gates	
Pre_norm	Part of FPU	1877	8-bit reduced OR → AND	29.4	50.8	1	19/19	83/83	0.4
MD5	MD5 full chip	13111	Incorrect state transition	5294	5670	2	33/64	58/126	28.2
DLX	5-stage pipeline MIPS-Lite CPU	14725	JAL inst. leads to incorrect bypass from MEM stage	25674	78834	54	1/21	1/944	1745
			Incorrect inst. forwarding	29436	30213	6	1/2	1/2	85

DPS is used in this experiment. The error-diagnosis technique is based on [6]. "#Supp." is the number of support signals and "#Gates" is the number of gates in the resynthesized netlist. The numbers are shown as minimum/maximum.

The comparison between the first and total diagnosis time in Table 9.5 shows that diagnosing the first trace takes more than 30% of the total diagnosis time in all the benchmarks. The reason is that the first diagnosis can often localize errors to a small number of sites, which reduces the search space of further diagnoses significantly. Since CoRé relies on iterative diagnosis to refine the abstraction of signatures, this phenomenon ensures that CoRé is efficient after the first iteration. As Table 9.5 shows, error diagnosis is still the bottleneck of the CoRé framework. We also observe that fixing some bugs requires a large number of gates and support variables in their resynthesized netlists because the bugs are complex functional errors injected at the RTL.

9.3.2 RTL Error Repair

In RTL error-repair experiments, we evaluated the performance of the techniques described in Section 9.2 with a range of Verilog benchmarks. We used a proprietary Perl-based Verilog parser to insert conditional assignments into RTL code. Synthesis-based diagnosis was implemented using OpenAccess 2.2 and OAGear 0.96 [162] with RTL Compiler v4.10 from Cadence as the synthesis tool. For simulation-based diagnosis, we adopted an RTL symbolic simulator, Insight 1.4, from Avery Design Systems [147]. For efficiency, we implemented the techniques described in [62] to convert PB problems to SAT problems and adopted MiniSat as our SAT solver [61]. All the experiments were conducted on an AMD Opteron 880 (2.4GHz) Linux workstation with 16GB memory. The designs under test included several circuits selected from OpenCores [154] (Pre_norm, MD5, MiniRISC, and CF_FFT), the picoJava-II microprocessor (Pipe), DLX, and Alpha. Bugs (described in Table 9.6) were injected into these benchmarks, with the exception of DLX and Alpha, which already included bugs. We used constrained-random simulation to generate bug traces for Pipe, Pre_norm, and CF_FFT, while the bug traces for the rest of the benchmarks were generated using the verification environment shipped with the designs. Traces to expose bugs in DLX and Alpha were given by the verification engineer and were generated using a constrained-random simulation tool [132]. The number of traces for the benchmarks and their lengths are also reported in Table 9.6. The characteristics of these benchmarks are summarized in Table 9.7. In the table, "RTL #Lines" is the number of lines of RTL code in a design, and "Gate-level #Cells" is the cell count of the synthesized netlist. To compare our results with previous work, we implemented the algorithms for gate-level error diagnosis in [6, 125]. In the table, we list the number of MUXes inserted by their techniques in column "#MUXes", and the number of conditional assignments under "#Assi.".

Synthesis-Based Error Diagnosis

In this experiment, we performed combinational and sequential error diagnosis using the synthesis-based techniques described in Section 9.2.2. For comparison with previous work, we also synthesized the benchmarks and performed gate-level error diagnosis using Smith's and Ali's [6, 125] techniques described in Section 5.1.3. The results are summarized in Table 9.8 and Table 9.9. Recall that a *symptom core* suggests a possible set of signals to modify for correcting the design, and it includes one or more *symptom variables*. In all our experiments, we found that the reported symptom cores included the root causes of errors for all benchmarks. In other words, REDIR accurately pointed out the signals that exhibited incorrect behavior.

Comparison between RTL and gate-level error diagnosis: this comparison clearly indicates that diagnosing functional errors at the RTL has significant advantages over the gate level, including shorter runtime and more accurate diagnoses. As Table 9.8 shows, most errors can be diagnosed using our techniques within a few minutes, while Table 9.9 shows that identifying the same errors at the gate level takes more than 48 hours in many cases. One major reason for this is that the number of possible symptom variables (error sites), i.e., internal netlist signals responsible for the bug, is significantly smaller in RTL diagnosis, as can be observed from the numbers of inserted conditional assignments shown in Table 9.7. This is due to the fact that one simple RTL statement may be synthesized into a complex netlist, which proliferates the

Table 9.6. Description of bugs in benchmarks.

Bench-mark	Bug ID	Description	Bug traces	
			#Traces	#Cycles
Pipe	A	One signal inverted	32	200
Pre_norm	A	Reduced OR replaced by reduced AND	32	20
	B	One signal inverted	32	20
	C	One 26-bit bus MUX select line inverted	32	20
	D	Bug A + Bug B	32	20
	E	Bug A + Bug B + Bug C	32	20
MD5	A	Incorrect operand for a 32-bit addition	1	200
	B	Incorrect state transition	1	200
	C	Bug B with a shorter trace	1	50
MRISC	A	Incorrect RHS for a 11-bit value assignment	1	200
CF_FFT	A	One signal inverted	32	15
DLX	A	SLL inst. does shift the wrong way	1	150
	B	SLTIU inst. selects the wrong ALU operation	1	68(178)
	C	JAL inst. leads to incorrect bypass from MEM stage	1	47(142)
	D	Incorrect forwarding for ALU+IMM inst.	1	77(798)
	E	Does not write to reg31	1	49(143)
	F	RT reads lower 30 bits only	1	188
	G	If RT = 7 memory write is incorrect	1	30(1080)
Alpha	A	Write to zero-reg succeeds if rdb_idx = 5	1	70(256)
	B	Forwarding through zero reg on rb	1	83(1433)
	C	Squash if source of MEM/WB = dest. of ID/EX and instr. in ID is not a branch	1	150(9950)

DLX and Alpha included native bugs, while other bugs were manually injected. Bug traces for several DLX and Alpha benchmarks have been minimized before diagnosis, and their original lengths are shown in parentheses.

number of error sites. For example, a statement like "a = b + c" creates only one symptom variable at the RTL. Its synthesized netlist, however, may contain hundreds of error sites, depending on the implementation of the adder and the bit-width of the signals. The small number of potential symptom variables at the RTL significantly reduces the search space for PB or SAT solvers and provides very short diagnosis runtime. In addition, one bug at the RTL may transform into multiple simultaneous bugs at the gate level. Since runtime of error diagnosis grows substantially with each additional bug [125], being able to diagnose errors at the RTL avoids the expensive multi-error diagnosis process at the gate level. We also observed that although the runtime of the RTL error diagnosis still increases with each additional bug, its growth rate is much smaller than the growth rate at the gate level. For example, as Table 9.9 shows, the runtime of the gate-level diagnosis for Pre_norm(A) and (D), which combined (A) and (B), was 63.6 and 88.7 seconds, respectively. On the other hand, Table 9.8 shows that the runtime for RTL diagnosis was13.2 and 13.8 seconds,

Table 9.7. Characteristics of benchmarks.

Bench-mark	Description	#Flip-flops	Trace type	Gate-level [6, 125]		RTL (Ours)	
				#Cells	#MUXes	#Lines	#Assi.
Pipe	Part of PicoJava pipeline control unit	2	Constrained-random	55	72	264	31
Pre_norm	Part of FPU	71	Constrained random	1877	1877	270	43
MD5	MD5 full chip	910	Direct test	13311	13313	438	37
MiniRISC	MiniRISC full chip	887	Direct test	6402	6402	2013	43
CF_FFT	Part of the CF_FFT chip	16,638	Constrained-random	126532	126560	998	223
DLX	5-stage pipeline CPU, MIPS-Lite ISA	2,062	Constrained-random	14725	14727	1225	84
Alpha	5-stage pipeline CPU, Alpha ISA	2,917	Constrained-random	38299	38601	1841	134

"#MUXes" is the number of MUXes inserted by gate-level diagnosis [6, 125] for comparison, and "#Assi." is the number of conditional assignments inserted by our solution.

respectively. These results clearly indicate that adopting gate-level techniques into RTL is the correct approach: it provides excellent accuracy because formal analysis can be performed, yet it avoids drawbacks in gate-level analysis in that it is still highly scalable and efficient. This is achieved by our new constructs that model errors at the RTL instead of the gate level. These results also demonstrate that trying to diagnose RTL errors at the gate level and mapping the results back to the RTL is ineffective and inefficient, not to mention the fact that such a mapping is usually difficult to find.

Comparison between combinational and sequential diagnosis: the difference between combinational and sequential diagnosis is that sequential diagnosis only uses output responses for constraints, while combinational is allowed to use state values. As Table 9.8 shows, the runtime of combinational diagnosis is typically shorter, and the number of symptom cores is often smaller. In DLX(D), for example, the combinational technique runs significantly faster than sequential, and returns only three cores, while sequential returns nine. The reason is that combinational diagnosis allows the use of state values, which provide additional constraints to the PB instance. As a result, the PB solver can find solutions faster, and the additional constraints further localize the bugs. Being able to utilize state values is especially important for designs with very deep pipelines, where an error may be observed hundred cycles later. For example, the error injected into CF_FFT requires more than 40 cycles to propagate to any primaryoutput, making the use of sequential diagnosis difficult. In

Table 9.8. RTL synthesis-based error-diagnosis results.

Bench-mark	Bug ID	RTL diagnosis (Our work)					
		Combinational			Sequential		
		Errors found		Runtime	Errors found		Runtime
		#Symp.	#Cores	(sec)	#Symp.	#Cores.	(sec)
Pipe	A	1	1	6.0	1	1	**6.0**
Pre_ norm	A	1	1	13.2	1	1	**13.2**
	B	1	1	11.4	1	2	**13.4**
	C	1	1	11.4	1	1	**11.4**
	D	2	1	12.4	2	2	**13.8**
	E	3	2	13.9	3	4	**17.4**
MD5	A	1	1	83.3	1	3	**173.2**
	B	1	1	42.9	1	2	**110.1**
	C	1	1	14.1	1	6	**49.8**
MRISC	A	States unavailable			1	2	**32.0**
CF_FFT	A	1	4	364.8	Trace unavailable		
DLX	A	1	1	41.2	1	3	**220.8**
	B	1	4	54.8	1	17	**1886.3**
	C	1	5	15.8	1	11	**104.0**
	D	1	3	27.5	1	9	**2765.1**
	E	1	4	19.1	1	12	**105.2**
	F	1	2	67.8	1	2	**457.4**
	G	1	1	11.3	Trace unavailable		
Alpha	A	1	5	127.4	1	9	**525.3**
	B	1	5	111.6	1	5	**368.9**
	C	1	3	122.3	1	3	**250.5**

"#Symp." is the number of *symptom variables* in each core, and "#Cores" is the total number of *symptom cores*. The results should be compared with Table 9.9, which show that RTL diagnosis outperforms gate-level diagnosis in all the benchmarks: the runtime is shorter, and the diagnosis is more accurate.

addition, bugs that are observed in design states can only be diagnosed when state values are available, such as DLX(G). On the other hand, sequential diagnosis is important when state values are unavailable. For example, the bug injected into the MiniRISC processor changed the state registers, damaging correct state values. In practice, it is also common that only responses at primary outputs are known. Therefore, being able to diagnose errors in combinational and sequential circuits is equally important, and both are supported by REDIR.

The comparison between MD5(B) and MD5(C) shows that there is a trade-off between diagnosis runtime and quality: MD5(C) uses a shorter trace and thus requires shorter diagnosis runtime; however, the number of symptom coresis larger than that returned by MD5(B), showing that the results are less

Table 9.9. Gate-level error-diagnosis results.

Bench-mark	Bug ID	Gate-level diagnosis [6, 125]					
		Combinational			Sequential		
		Errors found		Runtime	Errors found		Runtime
		#Sites	#Cores	(sec)	#Sites	#Cores	(sec)
Pipe	A	1	1	6.9	1	1	**7.1**
Pre_norm	A	1	1	51.1	1	1	**63.6**
	B	1	3	41.6	1	4	**46.7**
	C	**Time-out (48 hours) with > 10 error sites**					
	D	2	3	73.3	2	4	**88.7**
	E	**Time-out (48 hours) with > 8 error sites**					
MD5	A	**Time-out (48 hours) with > 6 error sites**					
	B	1	2	10980	1	4	**41043**
	C	1	3	2731	1	28	**17974**
MRISC	A	States unavailable			**Time-out (48 hours)**		
CF_FFT	A	1	1	109305	Trace unavailable		
DLX	A	**Time-out (48 hours)**			**Out of memory**		
	B	1	20	15261	**Out of memory**		
	C	1	45	11436	1	170	**34829**
	D	1	6	18376	1	6	**49787**
	E	1	12	9743.5	1	193	**19621**
	F	1	10	15184	**Out of memory**		
	G	1	9	4160.1	Trace unavailable		
Alpha	A	**Time-out (48 hours)**					
	B	**Time-out (48 hours)**					
	C	**Out of memory**					

"#Sites" is the number of error sites reported in each core, and "#Cores" is the total number of symptom cores returned by error diagnosis.

accurate. The reason is that longer traces usually contain more information; therefore, they can better localize design errors. One way to obtain short yet high-quality traces is to perform bug trace minimization before error diagnosis. Such minimization techniques can remove redundant information from the bug trace and greatly facilitate error diagnosis. We used the Butramin technique described in Chapter 8 to minimize the traces for DLX and Alpha, and the length of the original traces is shown in parentheses. In general, one trace is enough to localize the errors to a small number of symptom cores, while additional traces may further reduce this number.

Case study: we use DLX(D) as an example to show the power of our error-diagnosis techniques. Part of its RTL code is shown below:

```
always@(memstage or exstage or idstage or rs3rd or rs3rt or rs4rd or rs4rt or rsr31)
 casex ({memstage,exstage,idstage,rs3rd,rs3rt,rs4rd,rs4rt,rsr31})
  {'ALUimm, 'dc3, 'dc3,'dc,'dc, 'dc, 'true,'dc}:
  RSsel = 'select_stage3_bypass; // Buggy
 ......
```

In this example, the buggy code selects stage3 bypass, while the correct implementation should select stage4. Error diagnosis returns two *symptom cores*: *RSsel* and *ALUout*. Obviously, *RSsel* is the correct diagnosis. However, *ALUout* is also a correct diagnosis because if the ALU can generate correct outputs even though the control signal is incorrect, then the bug can also be fixed. Nonetheless, this is not a desirable fix. This case study shows that REDIR can suggest various ways to repair the same error, allowing the designer to consider different possibilities in order to choose the best fix.

Simulation-Based Error Diagnosis

In this experiment, we performed simulation-based diagnosis using the algorithm described in Section 9.2.2 with Insight, an RTL symbolic simulator from [147]. Benchmarks Pipe and CF_FFT were used in this experiment. Simulation took 23.8 and 162.9 seconds to generate SAT instances for these benchmarks, respectively. The SAT solver included in Insight then solved the instances in 1 and 723 seconds respectively, and it successfully identified the design errors. Note that currently, the SAT solver only returns one, instead of all possible symptom cores. Although the runtime of simulation-based approach is longer than the synthesis-based method, it does not require the design to be synthesized in advance, thus saving the synthesizer runtime.

Error Correction

In our error-correction experiment, we applied the techniques described in Section 9.2.3 to fix the errors diagnosed in Table 9.8. We used combinational diagnosis in this experiment, and corrected the error locations using the resynthesis methods described in Chapter 6. We summarize the results in Table 9.10 where we indicate which of the two synthesis techniques we used, either GDS or DPS. In the table, "#Cores fixed" is the number of symptom cores that can be corrected using our error-correction techniques, and "#Fixes" is the number of ways to fix the errors. We applied GDS first in the experiment, and observed that GDS often returns a large number of valid fixes that can correct the design errors. One reason is that GDS performs exhaustive search to find new logic expressions; therefore, it may find many different ways to produce the same signal. For example, "$A \cdot \overline{B}$" and "$A \cdot (A \oplus B)$" are both returned even though they are equivalent. Another reason is that we only diagnosed short bug traces, which may produce spurious fixes: signatures of different variables are the same even though their functions are different. As a result, we only report the first 100 fixes in our implementation, where the fixes are sorted so that those with smaller number of logic operations are returned first. Due to the exhaustive-search nature of GDS, memory usage of GDS may be high during the search, as are the cases for benchmarks DLX (C–F) and Alpha. In these benchmarks, GDS ran out of memory, and we relied on DPS to find fixes that can correct the errors. Since DPS only returns one logic expression when fixing an error, the number of possible fixes is significantly smaller.

Table 9.10. Error-correction results for RTL designs

Bench-mark mark	Bug ID	#Cores fixed	Resyn. method	#Fixes	Runtime (sec)
Pipe	A	1	GDS	2214	1.0
Pre_norm	A	1	GDS	4091	1.1
	B	1	GDS	4947	2.4
	C	1	GDS	68416	5.6
	D	2	GDS	79358	7.1
	E	3	GDS	548037	41.6
MD5	A	1	GDS	33625	4.1
	B	0	GDS	0	3.86
CF_FFT	A	3	GDS	214800	141.6
DLX	A	0	GDS	0	1.3
	B	3	GDS	5319430	111.2
	C	5	DPS	5	1.6
	D	3	DPS	3	1.6
	E	4	DPS	4	1.4
	F	2	DPS	2	2.9
	G	1	GDS	51330	0.7
Alpha	A	5	DPS	5	7.9
	B	4	DPS	4	10.4
	C	3	DPS	3	8.5

Table 9.10 shows that we could not find valid fixes for benchmarks MD5(B) and DLX(A). The reason is that the bugs in these benchmarks involve multi-bit variables. For example, bug MD5(b) is an incorrect state transition for a 3-bit state register. Since in this experiment we only consider the least-significant bits of such variables during error correction, we could not find a valid fix. This problem can be solved by inserting a conditional assignment for every bit in a multi-bit variable.

Discussion of RTL Error-Repair Results

The RTL error-diagnosis results show that our error-modeling constructs and diagnosis techniques can effectively localize design errors to a small number of symptom variables. On the other hand, our error-correction results suggest that options to repair the diagnosed errors abound. The reason is that the search space of error correction is much larger than error diagnosis: there are various ways to synthesize a logic function. As a result, finding high-quality fixes for a bug requires much more information than providing high-quality diagnoses. Although this can be achieved by diagnosing longer or more numerous bug traces, the runtime of REDIR will also increase.

This observation shows that automatic error correction is a much more difficult problem than automatic error diagnosis. In practice, however, engineers often find error diagnosis more difficult than error correction. It is common that engineers need to spend days or weeks finding the cause of a bug. However, once the bug is identified, fixing it may only take a few hours. To this end, our error-correction technique can also be used to facilitate manual error repair, and it works as follows: (1) the engineer fixes the RTL code manually to provide new logic functions for the symptom cores identified by error diagnosis; and (2) REDIR simulates the new functions to check whether the signatures of symptom cores can be generated correctly using the new functions. If the signatures cannot be generated by the new functions, then the fix is invalid. In this way, engineers can check the correctness of their fixes before running verification, which can accelerate the manual error-repair process significantly.

The synthesis-based results show that our techniques can effectively handle designs as large as 2000 lines of RTL code, which is approximately the size that an engineer actively works on. Since synthesis tools are available in most companies, REDIR can be used by engineers everyday to facilitate their debugging process. On the other hand, the simulation-based results suggest that our techniques are promising. Once RTL symbolic simulators become accessible to most companies, REDIR can automatically exploit their simulation power to handle even larger designs.

9.4 Summary

In this chapter we empirically evaluated the effectiveness of the CoRé framework in repairing functional errors in combinational gate-level netlists. In addition, we extended the framework to repair errors in sequential circuits. This framework exploits both satisfiability and observability don't-cares, and it uses an abstraction-refinement scheme to achieve better scalability. The experimental results show that CoRé can produce a modified netlist which eliminates erroneous responses while maintaining correct ones. In addition, CoRé only requires test vectors and correct output responses; therefore, it can be easily adopted in most verification flows.

Other key ideas presented in this chapter are the constructs and algorithms that provide a new way to diagnose and correct errors at the RTL, including: (1) an RTL error modeling construct; (2) scalable error-diagnosis algorithms using Pseudo-Boolean constraints, synthesis, and simulation; and (3) an error-correction technique using signatures. To empirically validate our proposed techniques, we developed a new verification framework, called REDIR. To this end, our experiments with industrial designs demonstrate that REDIR is efficient and scalable. In particular, designs up to a few thousand lines of code (or 100K cells after synthesis) can be diagnosed within minutes with high accuracy.

The comparison between gate-level and RTL error diagnosis shows that RTL bugs should be fixed at the RTL because fixing the same errors at the gate level will become much more difficult. To this end, REDIR can greatly enhance the RTL debugging process to prevent bugs from escaping to the gate level, allowing most functional errors to be caught and repaired at the RTL. Therefore, even if bugs still escape to the gate level, those bugs will be more subtle and should require smaller changes to the netlist. This will allow gate-level error-repair techniques to work more effectively.

Chapter 10

INCREMENTAL VERIFICATION FOR PHYSICAL SYNTHESIS

As interconnect increasingly dominates delay and power at the latest technology nodes, much effort is invested in physical synthesis optimizations, posing great challenges in validating the correctness of such optimizations. Common design methodologies that delay the verification of physical synthesis transformations until the completion of the design phase are no longer sustainable because it makes the isolation of potential errors extremely challenging. Since the design's functional correctness should not be compromised, engineers dedicate considerable resources to ensure the correctness at the expense of improving other aspects of design quality. To address these challenges, we propose a fast incremental verification system for physical synthesis optimizations, called InVerS, which includes capabilities for error detection and diagnosis. This system helps engineers discover errors earlier, which simplifies error isolation and correction, thereby reducing verification effort and enabling more aggressive optimizations to improve performance.

10.1 Background

In this section we first take a closer look at the current physical synthesis flow. Next, we describe a powerful physical synthesis technique called retiming. Retiming repositions registers in a design and can perform optimizations not achievable by combinational resynthesis methods. Our methodologies to verify the correctness of these optimizations will be presented in the next section.

10.1.1 The Current Physical Synthesis Flow

Post-placement optimizations have been studied and used extensively to improve circuit parameters such as power and timing, and these techniques are often called physical synthesis. In addition, it is sometimes necessary to change the layout manually in order to fix bugs or optimize specific objectives; this process is called Engineering Change Order (ECO). Physical synthesis is commonly performed using the following flow: (1) perform accurate analysis of the optimization objective, (2) select gates to form a region for optimization, (3) resynthesize the region to optimize the objective, and (4) perform legalization to repair the layout. The work by Lu et al. [95] and Changfan et al. [48] are all based on this flow.

Given that subtle and unexpected bugs still appear in physical synthesis tools today [9], verification must be performed to ensure the correctness of the circuit. However, verification is typically slow; therefore, it is often performed after hundreds or thousands of optimizations, as shown in Figure 10.1. As a result, it is difficult to identify the circuit modification that introduced the bug. In addition, debugging the circuit at this design stage is often difficult because engineers are unfamiliar with the automatically generated netlist. As we will show later, InVerS addresses these problems by providing a fast incremental verification technique.

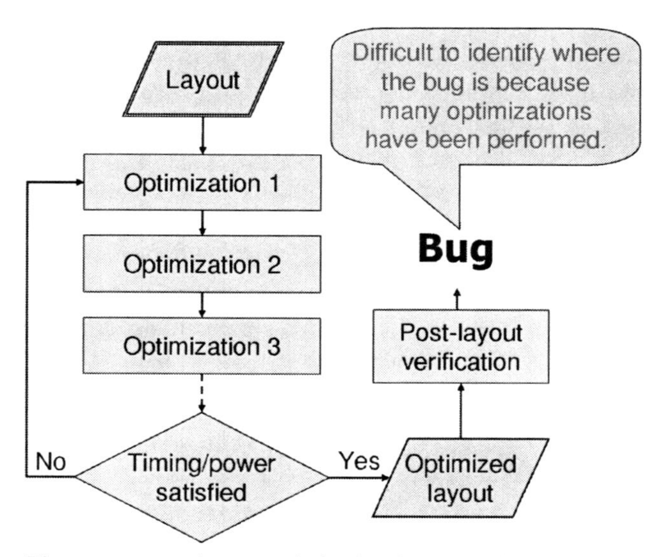

Figure 10.1. The current post-layout optimization flow. Verification is performed after the layout has undergone a large number of optimizations, which makes debugging difficult.

10.1.2 Retiming

Retiming is a sequential logic optimization technique that repositions the registers in a circuit while leaving the combinational cells unchanged [88, 120]. It is often used to minimize the number of registers in a design or to reduce a circuit's delay. For example, the circuit in Figure 10.4(b) is a retimed version of the circuit in Figure 10.4(a) that optimizes delay. Although retiming is a powerful technique, ensuring its correctness imposes a serious problem on verification because sequential equivalence checking is orders of magnitude more difficult than combinational equivalence checking [74]. As a result, the runtime of sequential verification is often much longer than that of combinational verification, if it ever finishes. This problem will be addressed in Section 10.2.2.

10.2 Incremental Verification

We provide a robust incremental verification package that is composed of a logic simulator, a SAT-based formal equivalence checker, and our innovative similarity metric between a circuit and its revision. In this section we define our similarity metrics, and then describe our overall verification methodology.

10.2.1 New Metric: Similarity Factor

We define the *similarity factor* as an estimate of the similarity between two netlists, ckt_1 and ckt_2. This metric is based on simulation signatures of individual signals, and those signatures can be calculated using fast simulation. Let N be the total number of signals (wires) in both circuits. Out of those N signals, we distinguish M *matching* signals – a signal is considered matching if and only if both circuits include signals with an identical signature. The similarity factor between ckt_1 and ckt_2 is then M/N. In other words:

$$similarity\ factor = \frac{number\ of\ matching\ signals}{total\ number\ of\ signals} \tag{10.1}$$

We also define the *difference factor* as $(1 - similarity\ factor)$.

EXAMPLE 10.1 *Consider the two netlists shown in Figure 10.2, where the signatures are shown above the wires. There are 10 signals in the netlists, and 7 of them are matching. As a result, the similarity factor is 7/10= 70%, and the difference factor is 1 – 7/10 = 30%.*

Intuitively, the similarity factor of two identical circuits should be 100%. If a circuit is changed slightly but is still mostly equivalent to the original version, then its similarity factor should drop only slightly. For example, Figure 10.3(a) shows a netlist where a region of gates is resynthesized correctly. Since only the signatures in that region will be affected, the similarity factor only drops slightly. However, if the change greatly affects the circuit's function, the similarity factor can drop significantly, depending on the number of signals affected by the change. As Figure 10.3(b) shows, when a bug is introduced by resynthesis, the signatures in the output cone of the resynthesized region will also be different, causing a larger drop in similarity factor. However, two equivalent

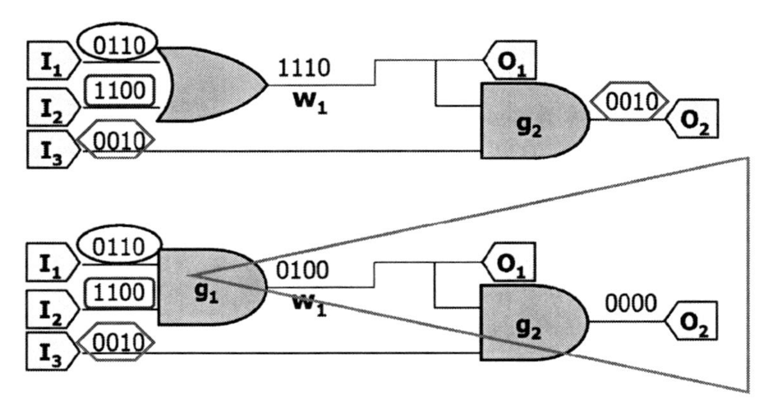

Figure 10.2. Similarity factor example. Note that the signatures in the fanout cone of the corrupted signal are different.

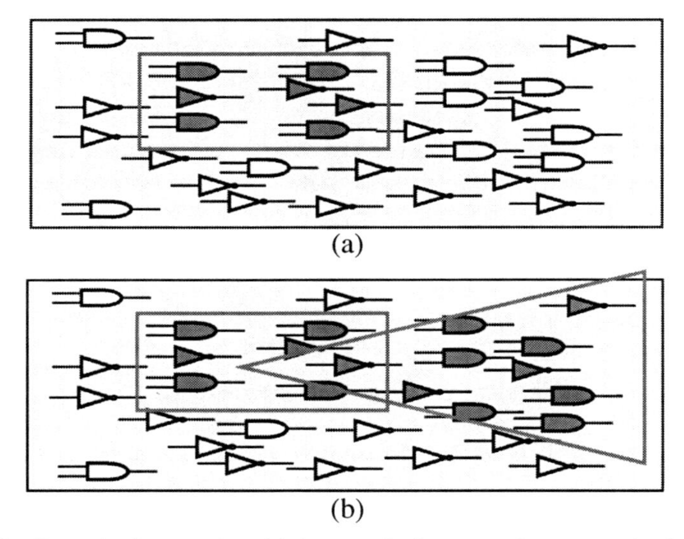

Figure 10.3. Resynthesis examples: (**a**) the gates in the rectangle are resynthesized correctly, and their signatures may be different from the original netlist; (**b**) an error is introduced during resynthesis, and the signatures in the output cone of the resynthesized region are also different, causing a significant drop in similarity factor.

circuits may be dissimilar, e.g., a Carry-Look-Ahead adder and a Kogge-Stone adder. Therefore, the similarity factor should be used in incremental verification and cannot replace traditional verification techniques.

10.2.2 Verification of Retiming

A signature represents a fraction of a signal's truth table, which in turn describes the information flow within a circuit. While retiming may change the clock cycle that certain signatures are generated, because combinational cells are preserved, most generated signatures should be identical. Figure 10.4 shows a retiming example from [44], where (a) is the original circuit and (b) is the retimed circuit. A comparison of signatures between the circuits shows that the signatures in (a) also appear in (b), although the cycles in which they appear may be different. For example, the signatures of wire w (boldfaced) in the retimed circuit appear one cycle earlier than those in the original circuit because the registers were moved later in the circuit. Otherwise, the signatures of (a) and (b) are identical. This phenomenon becomes more obvious when the circuit is unrolled, as shown in Figure 10.5. Since the maximum absolute lag in this example is 1, retiming only affects gates in the first and the last cycles, leaving the rest of the circuits identical. As a result, signatures generated by the unaffected gates should also be identical. Based on this observation, we extend our similarity factor to sequential verification, called *sequential similarity factor*, as follows. Assume two netlists, ckt_1 and ckt_2, where the total number of signals (wires) in both circuits is N. After simulating C cycles, $N \times C$ signatures will be generated. Out of those signatures, we distinguish M *matching* signatures. The sequential similarity factor between ckt_1 and ckt_2 is then $M/(N \times C)$. In other words:

$$sequential\ similarity\ factor = \frac{number\ of\ matching\ signatures\ for\ all\ cycles}{total\ number\ of\ signatures\ for\ all\ cycles} \tag{10.2}$$

10.2.3 Overall Verification Methodology

As mentioned in Section 10.1.1, traditional verification is typically performed after a batch of circuit modifications because it is very demanding and time consuming. As a result, once a bug is found, it is often difficult to isolate the change that introduced the bug because hundreds or thousands of changes have been made. Similarity factor addresses this problem by pointing out the changes that might have corrupted the circuit. As described in previous subsections, a change that greatly affects the circuit's function will probably cause a sudden drop in the similarity factor. By monitoring the change in similarity factor after every circuit modification, engineers will be able to know when a bug might have been introduced and traditional verification should be performed. Using the techniques that we developed, we propose the InVerS incremental verification methodology as shown in Figure 10.6, and it works as follows:

1 After each change to the circuit, the similarity factor between the new and the original circuit is calculated. Running average and standard deviation of the past 30 similarity factors are used to determine whether the current similarity factor has dropped significantly. Empirically, we found that if the

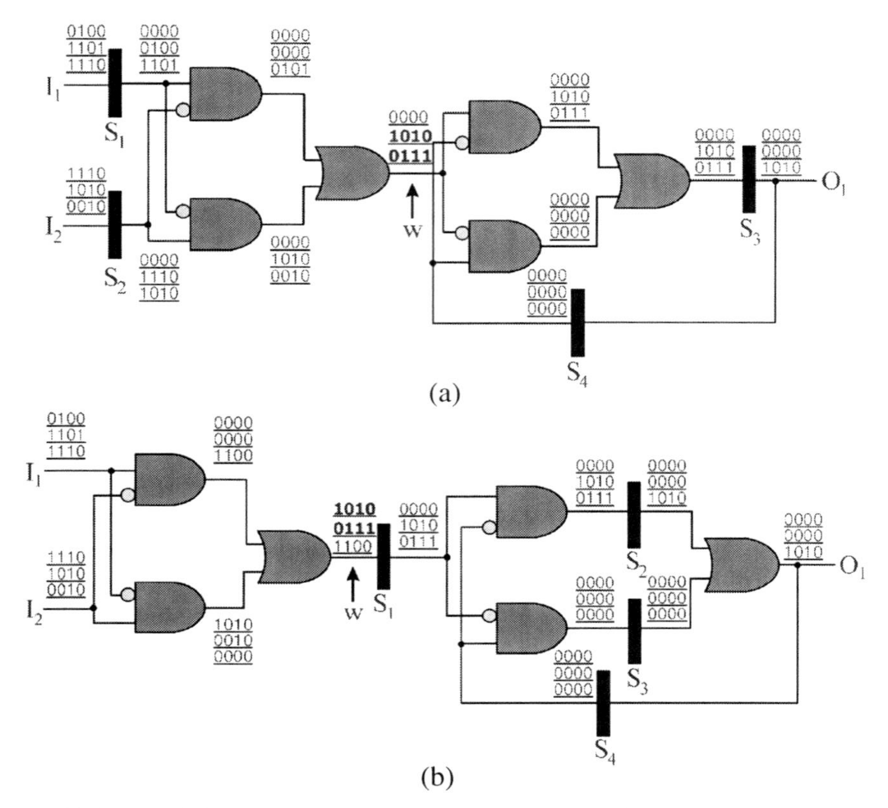

Figure 10.4. A retiming example: (**a**) is the original circuit, and (**b**) is its retimed version. The tables above the wires show their signatures, where the nth row is for the nth cycle. Four traces are used to generate the signatures, producing four bits per signature. Registers are represented by black rectangles, and their initial states are 0. As wire w shows, retiming may change the cycle that signatures appear, but it does not change the signatures (signatures shown in boldface are identical).

current similarity factor drops below the average by more than two standard deviations, then it is likely that the change introduced a bug. This number, however, may vary among different benchmarks and should be empirically determined.

2 When similarity factor indicates a potential problem, traditional verification should be performed to verify the correctness of the executed circuit modification.

3 If verification fails, our functional error repair tools can be used to repair the errors.

Since InVerS monitors drops in similarity factors, rather than the absolute values of similarity factors, the structures of the netlists become less relevant.

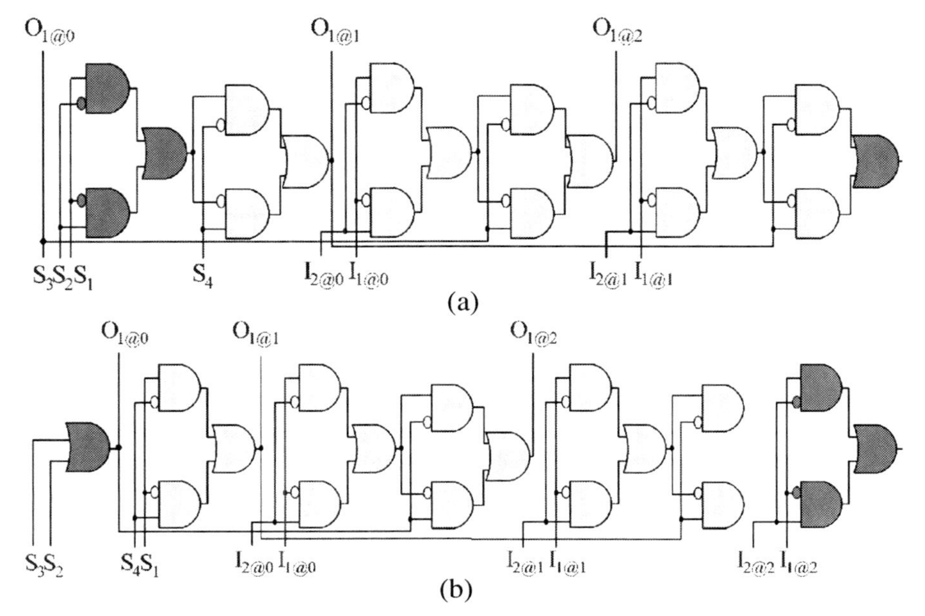

Figure 10.5. Circuits in Figure 10.4 unrolled three times. The cycle in which a signal appears is denoted using subscript "@". Retiming affects gates in the first and the last cycles (*marked in dark gray*), while the rest of the gates are structurally identical (*marked in light gray*). Therefore, only the signatures of the dark-gray gates will be different.

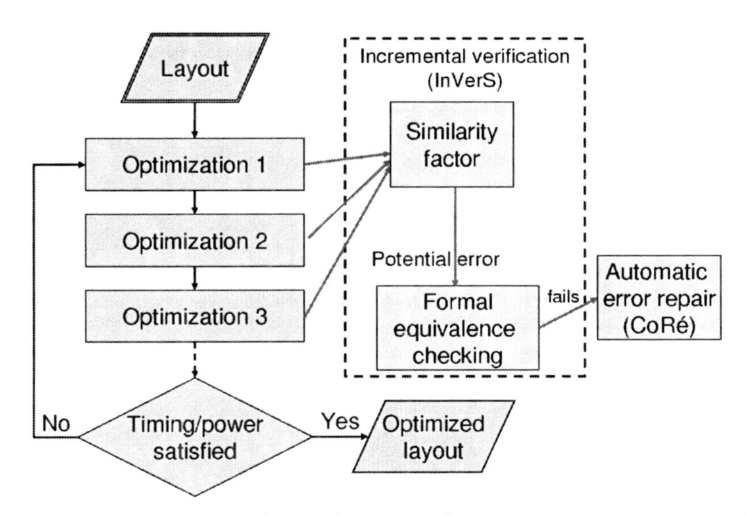

Figure 10.6. Our InVerS verification methodology. It monitors every layout modification to identify potential errors and calls equivalence checking when necessary. Our functional error repair techniques can be used to correct the errors when verification fails.

Therefore, InVerS can be applied to a variety of netlists, potentially with different error-flagging thresholds. As Section 10.3 shows, the similarity factor exhibits high accuracy for various practical designs and allows our verification methodology to achieve significant speed-up over traditional techniques.

10.3 Experimental Results

We implemented InVerS using OpenAccess 2.2 and OAGear 0.96 [162]. Our testcases were selected from IWLS'05 benchmarks [161] based on designs from ISCAS'89 and OpenCores suites, whose characteristics are summarized in Table 10.1. In the table, the average level of logic is calculated by averaging the logic level of 30 randomly selected gates. The number of levels of logic can be used as an indication of the circuit's complexity. We conducted all our experiments on an AMD Opteron 880 Linux workstation. The resynthesis package used in our experiments is ABC from UC Berkeley [148]. In this section we report results on combinational and sequential verification, respectively.

Table 10.1. Characteristics of benchmarks.

Benchmark	Cell count	Ave. level of logic	Function
S1196	483	6.8	ISCAS'89
USB_phy	546	4.7	USB 1.1 PHY
SASC	549	3.7	Simple asynchronous serial controller
S1494	643	6.5	ISCAS'89
I2C	1142	5.5	I2C master controller
DES_area	3132	15.1	DES cipher (area optimized)
SPI	3227	15.9	Serial parallel interface IP
TV80	7161	18.7	8-Bit microprocessor
MEM_ctrl	11440	10.1	WISHBONE memory controller
PCI_bridge32	16816	9.4	PCI bridge
AES_core	20795	11.0	AES cipher
WB_conmax	29034	8.9	WISHBONE Conmax IP core
DES_perf	98341	13.9	DES cipher (performance optimized)

10.3.1 Verification of Combinational Optimizations

Evaluation of the similarity factor: in our first experiment, we performed two types of circuit modifications to evaluate the effectiveness of the similarity factor for combinational verification. In the first type, we randomly injected an error into the circuit according to Abadir's error model (see Section 5.1.4), which includes the errors that occur frequently in gate-level netlists. This mimics the situation where a bug has been introduced. In the second type, we extracted a subcircuit from the benchmark, which was composed of 2–20 gates, and performed resynthesis of the subcircuit using ABC with the "resyn" command [148]. This is similar to the physical synthesis or ECO flow described in Section 10.1.1, where gates in a small region of the circuit are changed. We then calculated the similarity factor after each circuit modification for both types of circuit modifications and compared their difference. Thirty samples were used in this experiment, and the results are summarized in Table 10.2. From the results, we observe that both types of circuit modifications lead to decreases in similarity factor. However, the decrease is much more significant when an error is injected. As d_1 shows, the standardized differences in the means of most benchmarks are larger than 0.5, indicating that the differences are statistically significant. Since resynthesis tests represent the norm and error-injection tests are anomalies, we also calculated d_2 using only SD_r. As d_2 shows, the mean similarity factor drops more than two standard deviations when an error is injected for most benchmarks. This result shows that the similarity factor is effective in predicting whether a bug has been introduced by the circuit modification. Nonetheless, in all benchmarks, the maximum similarity factor for error-injection tests is larger than the minimum similarity factor for resynthesis tests, suggesting that the similarity factor cannot replace traditional verification and should be used as an auxiliary technique.

Table 10.2. Statistics of similarity factors for different types of circuit modifications.

Benchmark	Similarity factor (%)									
	Resynthesized				One error injected				d_1	d_2
	$Mean_r$	Min_r	Max_r	SD_r	$Mean_e$	Min_e	Max_e	SD_e		
USB_phy	99.849	99.019	100.000	0.231	98.897	91.897	99.822	1.734	0.969	4.128
SASC	99.765	99.119	100.000	0.234	97.995	90.291	99.912	2.941	1.115	7.567
I2C	99.840	99.486	100.000	0.172	99.695	98.583	100.000	0.339	0.567	0.843
SPI	99.906	99.604	100.000	0.097	99.692	96.430	99.985	0.726	0.518	2.191
TV80	99.956	99.791	100.000	0.050	99.432	94.978	100.000	1.077	0.930	10.425
MEM_ctrl	99.984	99.857	100.000	0.027	99.850	97.699	100.000	0.438	0.575	4.897
PCI_bridge32	99.978	99.941	100.000	0.019	99.903	97.649	99.997	0.426	0.338	3.878
AES_core	99.990	99.950	100.000	0.015	99.657	98.086	99.988	0.470	1.372	21.797
WB_conmax	99.984	99.960	100.000	0.012	99.920	99.216	99.998	0.180	0.671	5.184
DES_perf	99.997	99.993	100.000	0.002	99.942	99.734	100.000	0.072	1.481	23.969

Thirty tests were performed in this experiment, whose means, minimal values (Min), maximum values (Max), and standard deviations (SD) are shown. The last two columns show the standardized differences in the means: d_1 is calculated using the average of both SD_e and SD_r, while d_2 uses only SD_r.

The impact of cell count on the similarity factor: in order to study other aspects that may affect the similarity factor, we further analyze our results by plotting the factors against the cell counts of the benchmarks. To make the figure clearer, we plot the difference factor instead of the similarity factor. We notice that by construction, the difference factor tends to reduce with the increase in design size, which makes the comparison among different benchmarks difficult. In order to compensate this effect, we assume that the bug density is 1 bug per 1,000 gates and adjust our numbers accordingly. The plot is shown in Figure 10.7, where the triangles represent data points from error-injection tests, and the squares represent resynthesis tests. The linear regression lines of two data sets are also shown. From the figure, we observe that the difference factor tends to increase with the cell count for error-injection tests. The increase for resynthesis tests, however, is less significant. As a result, the difference factor of error-injected circuits (triangle data points) will grow faster than that of resynthesized circuits (square data points) when cell count increases, creating larger discrepancy between them. This result shows that the similarity factor will drop more significantly for larger designs, making it more accurate when applied to practical designs, which often have orders of magnitude more cells than the benchmarks used in our tests.

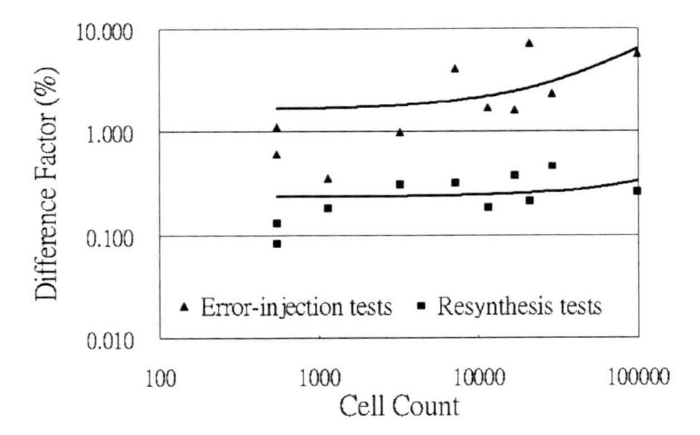

Figure 10.7. The relationship between cell count and the difference factor. The linear regression lines of the datapoints are also shown.

The impact of level of logic on the similarity factor: here we perform similar analysis using the number of levels of logic as the independent variable. The slopes of the linear regression lines for the error-injection tests and the resynthesis tests are 0.236 and 0.012, respectively. The difference in slopes shows that the difference factor grows faster when the number of levels of logic increases, indicating that the similarity factor will be more effective when designs become more complicated. This behavior is preferable because complicated designs are often more difficult to verify.

To study the impact of the number of levels of logic on the difference factor within a benchmark, we plotted the difference factor against the number of levels of logic using benchmark DES_perf in Figure 10.8. The logarithmic regression line for the error-injection tests are also shown. As the figure suggests, the difference factor decreases with the increase in the number of levels of logic. The reason is that gates with smaller numbers of levels of logic have larger downstream logic, therefore larger numbers of signatures will be affected. As a result, the difference factor will be larger. That the variance explained is large (0.7841) suggests that this relation is strong. However, some benchmarks do not exhibit this trend. For example, the variance explained for benchmark TV18 is only 0.1438. For benchmarks that exhibit this trend, the similarity factor provides a good predication of the location of the bug: a larger drop in the similarity factor indicates that the bug is closer to primary inputs.

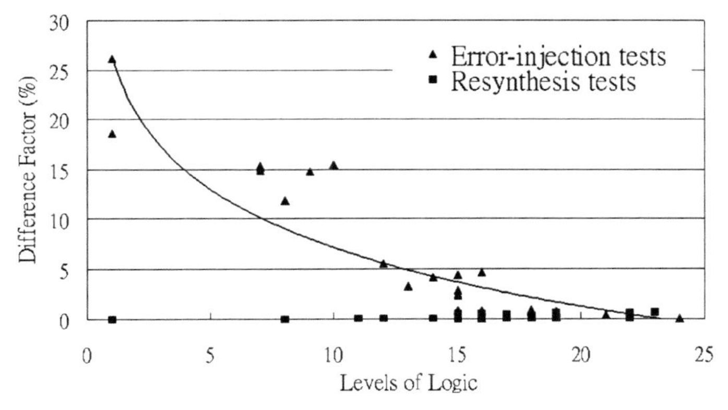

Figure 10.8. The relationship between the number of levels of logic and the difference factor in benchmark DES_perf. The x-axis is the level of logic that the circuit is modified. The logarithmic regression line for the error-injection tests is also shown.

To evaluate the effectiveness of our incremental verification methodology described in Section 10.2.3, we assumed that there is 1 bug per 100 circuit modifications, and then we calculated the accuracy of our methodology. We also report the runtime for calculating the similarity factor and the runtime for equivalence checking of each benchmark. Since most circuit modifications do not introduce bugs, we report the runtime when equivalence is maintained. The results are summarized in Table 10.3. From the results, we observe that our methodology has high accuracy for most benchmarks. In addition, the results show that calculating the similarity factor is significantly faster than performing equivalence checking. For example, calculating the similarity factor of the largest benchmark (DES_perf) takes less than 1 second, while performing equivalence checking takes about 78 minutes. Due to the high accuracy of the similarity factor, our incremental verification technique identifies more than

99% of errors, rendering equivalence checking unnecessary in those cases and providing a more than 100X speed-up.

Table 10.3. The accuracy of our incremental verification methodology.

Benchmark	Cell count	Accuracy (%)	Runtime(sec)	
			EC	SF
USB_phy	546	92.70	0.19	<0.01
SASC	549	89.47	0.29	<0.01
I2C	1142	95.87	0.54	<0.01
SPI	3227	96.20	6.90	<0.01
TV80	7161	96.27	276.87	0.01
MEM_ctrl	11440	99.20	56.85	0.03
PCI_bridge32	16816	99.17	518.87	0.04
AES_core	20795	99.33	163.88	0.04
WB_conmax	29034	92.57	951.01	0.06
DES_perf	98341	99.73	4721.77	0.19

1 bug per 100 circuit modifications is assumed in this experiment. Runtime for similarity-factor calculation (SF) and equivalence checking (EC) is also shown.

10.3.2 Sequential Verification of Retiming

In our second experiment, we implemented the retiming algorithm described in [88] and used our verification methodology to check the correctness of our implementation. This methodology successfully identified several bugs in our initial implementation. In our experience, most bugs were caused by incorrect netlist modifications when repositioning the registers, and a few bugs were due to erroneous initial state calculation. Examples of the bugs include: (1) incorrect fanout connection when inserting a register to a wire which already has a register; (2) missing/additional register; (3) missing wire when a register drives a primary output; and (4) incorrect state calculation when two or more registers are connected in a row.

To quantitatively evaluate our verification methodology, we ran each benchmark using the correct implementation and the buggy version to calculate their respective Sequential Similarity Factors (SSFs), where 10 cycles were simulated. The results are summarized in Table 10.4, which shows that the sequential similarity factors for retimed circuits are 100% for most benchmarks. As explained in Section 10.2.2, only a few signatures should be affected by retiming. Therefore, the drop in similarity factor should be very small, making sequential similarity factor especially accurate for verifying the correctness of retiming. This phenomenon can also be observed from Table 10.5, where the accuracy of our verification methodology is higher than 99% for most benchmarks. To compare our methodology with formal equivalence checking, we also show the runtime of a sequential equivalence checker based on bounded model checking in Table 10.5. This result shows that our methodology is more beneficial for sequential verification than combinational because sequential equivalence checking requires much more runtime than combinational. Since the runtime to compute sequential similarity factor remains small, our technique can still be applied after every retiming optimization and thus eliminating most unnecessary sequential equivalence checking calls.

Table 10.4. Statistics of sequential similarity factors for retiming with and without errors.

Benchmark	Sequential similarity factor (%)							
	Retiming without errors				Retiming with errors			
	$Mean_r$	Min_r	Max_r	SD_r	$Mean_e$	Min_e	Max_e	SD_e
S1196	100.0000	100.0000	100.0000	0.0000	98.3631	86.7901	100.0000	3.0271
USB_phy	100.0000	100.0000	100.0000	0.0000	99.9852	99.6441	100.0000	0.0664
SASC	99.9399	99.7433	100.0000	0.0717	99.9470	99.3812	100.0000	0.1305
S1494	100.0000	100.0000	100.0000	0.0000	99.0518	94.8166	99.5414	1.5548
I2C	100.0000	100.0000	100.0000	0.0000	99.9545	99.6568	100.0000	0.1074
DES_area	100.0000	100.0000	100.0000	0.0000	95.9460	69.1441	100.0000	6.3899

Thirty tests were performed in this experiment, whose means, minimal values (Min), maximum values (Max), and standard deviations (SD) are shown.

Table 10.5. Runtime of sequential similarity factor calculation (SSF) and sequential equivalence checking (SEC).

Benchmark	Cell count	DFF count	Accuracy (%)	Runtime (sec)	
				SEC	SSF
S1196	483	18	99.87	5.12	0.42
USB_phy	546	98	99.10	0.41	0.34
SASC	549	117	95.80	5.16	0.56
S1494	643	6	99.47	2.86	0.45
I2C	1142	128	99.27	2491.01	1.43
DES_area	3132	64	99.97	49382.20	14.50

Accuracy of our verification methodology is also reported, where 1 bug per 100 retiming optimizations is assumed.

10.4 Summary

In this chapter we presented a novel incremental equivalence verification system, InVerS, with a particular focus on improving design quality and engineers' productivity. The high performance of InVerS allows designers to invoke it frequently, possibly after each circuit transformation. This allows errors to be detected sooner, when they can be more easily pinpointed and resolved. The scalability of InVerS stems from the use of fast simulation, which can efficiently calculate our proposed *similarity factor* metric to spot potential differences between two versions of a design. The areas where we detect a low similarity are spots potentially hiding functional bugs that can be subjected to more expensive formal techniques. The experimental results show that InVerS achieves a hundred-fold runtime speed-up on large designs compared to traditional techniques for similar verification goals. Our methodology and algorithms promise to decrease the number of latent bugs released in future digital designs and to facilitate more aggressive performance optimizations, thus improving the quality of electronic design in several categories.

Chapter 11

POST-SILICON DEBUGGING AND LAYOUT REPAIR

Modern IC designs have reached unparalleled levels of overall complexity, and thorough verification is becoming more difficult. Furthermore, the verification problem is exacerbated by the highly competitive market which requires shorter time-to-market. As a result, design errors are more likely to escape verification in the early stage of the design flow and are found after layout has been finished; or even worse, after the chip has been taped-out. Needless to say, these errors must be fixed before the IC can reach the market. Fixing such errors is often costly, especially when the chip has been taped-out. The key to reduce this cost is to preserve as much previous effort spent on the design as possible. In this chapter we present error-repair techniques that support the post-silicon debugging methodology described in Section 4.4. However, these techniques can be applied to pre-silicon layout optimization or error repair as well.

As mentioned in Section 2.4, design errors that occur post-silicon can be functional or electrical, and various physical synthesis techniques may be used to fix such errors. However, there is no metric to measure the impact of a physical synthesis technique on the layout. In this chapter, we first define and explore the concepts of physical safeness and logical soundness to measure such an impact. We observe from this analysis that most existing physical synthesis techniques do not allow post-silicon metal fix, and those support metal fix have limited error-repair capabilities. Therefore, we propose a *SafeResynth* technique that is more powerful yet has little impact on a layout. The next section then describes how SafeResynth can be applied to repair post-silicon electrical errors. In addition, the section also illustrates our new functional and electrical error repair techniques. This chapter concludes with experimental results and a brief summary.

11.1 Physical Safeness and Logical Soundness

The concept of *physical safeness* is used to describe the impact of an optimization technique on the placement of a circuit. Physically safe techniques only allow legal changes to a given placement; therefore, accurate analysis such as timing and congestion can be performed. Such changes are safe because they can be rejected immediately if the layout is not improved. On the other hand, unsafe techniques allow changes that produce a temporarily illegal placement. As a result, their evaluation is delayed, and it is not possible to reliably decide if the change can be accepted or must be rejected until later. Therefore, the average quality of unsafe changes may be worse than that of accepted safe changes. In addition, other physical parameters, such as wirelength or via count, may be impacted by unsafe transformations.

Similar to physical safeness, *logical soundness* is used to describe the perturbation to the logic made by the optimization techniques. Techniques that do not change the logic usually do not require verification. Examples for this type of optimization include gate sizing and buffer insertion. Techniques that change the logic of the circuit may require verification to ensure their correctness. For example, optimizations based on reconnecting wires require verification because any bug in the optimization process may change the circuit's behavior. Since local changes to combinational logic can be verified easily using equivalence checking, they are considered logically sound. However, small changes to sequential logic often have global implications and are much more difficult to verify, therefore we do not classify them as logically sound techniques. These techniques include the insertion of clocked repeaters and the use of retiming.

11.1.1 Physically Safe Techniques

Symmetry-based rewiring is one of the few physical synthesis techniques that is physically safe in nature. As described in Chapter 7, it exploits symmetries in logic functions, looking for pin reconnections that improve the optimization objective. For example, the inputs to an AND gate can be swapped without changing its logic function. Since only wiring is changed in this technique, the placement is always preserved. An example of symmetry-based rewiring is given in Figure 11.1(a).

The advantage of physically safe techniques is that the effects of any change are immediately measurable, therefore the change can be accepted or rejected reliably. As a result, circuit parameters will not deteriorate after optimization and no timing convergence problem will occur. However, the improvement gained from these techniques is often limited because they cannot aggressively modify the logic or use larger-scale optimizations. For example, in [32] timing improvement measured before routing is typically less than 10%.

11.1.2 Physically Unsafe Techniques

Traditional physical synthesis techniques are physically unsafe because they create cell overlaps and thus prevent immediate evaluation of changes. Although some of these techniques can be applied in a safe way, they may lose their strength. It follows that existing physical synthesis tools usually rely on unsafe techniques, planning to correct potentially illegal changes after the optimization phase is complete. A classification of these techniques and their impact on logic are discussed below.

Gate sizing and buffer insertion are two important techniques that do not change the logic, as shown in Figure 11.1(b) and (d). Gate sizing chooses the size of the gates carefully so that signal delay in wires can be balanced with gate delay, and the gates have enough strength to drive the wires. Buffer insertion adds buffers to drive long wires. The work by Kannan et al. [80] is based on these techniques.

Gate relocation moves gates on critical paths to better locations and also does not change the logic. An example of gate relocation is given in Figure 11.1(c). Ajami et al. [4] utilize this technique by performing timing-driven placement with global routing information using the notion of movable Steiner points. They formulate the simultaneous placement and routing problem as a mathematical program. The program is then solved by Han-Powell method.

Gate replication is another technique that can improve circuit timing without changing the logic. As Figure 11.1(e) shows, by duplicating $g5$, the delay to $g1$ and $g9$ can be reduced. Hrkic et al. [71] proposed a placement-coupled approach based on such technique. Given a placed circuit, they first extract replication trees from the critical paths after timing analysis, and then they perform embedding and post-unification to determine the gates that should be duplicated as well as their locations. Since duplicated gates may overlap with existing gates, at the end of the process, timing-driven legalization is applied. Although their approach improves timing by 1–36%, it also increases route length by 2–28%.

Traditional rewiring techniques based on addition or removal of redundant wires are not physically safe. The basic idea is to add one or more redundant wires to make a target wire redundant so that it becomes removable. Since gates must be modified to reflect the changes in wires, cell overlaps may occur. The work in [47] utilizes this technique using an ATPG reasoning approach.

Optimization techniques discussed so far can be made physically safe by rejecting all changes that create new overlaps. For example, this would allow inserting buffers only in overlap-free sites. However, the prevailing practice for these and many other optimizations is to first allow overlaps and then call a legalizer to fix the overlaps. According to our definition, this is physically unsafe. In other words, depending on how many overlaps are introduced, how

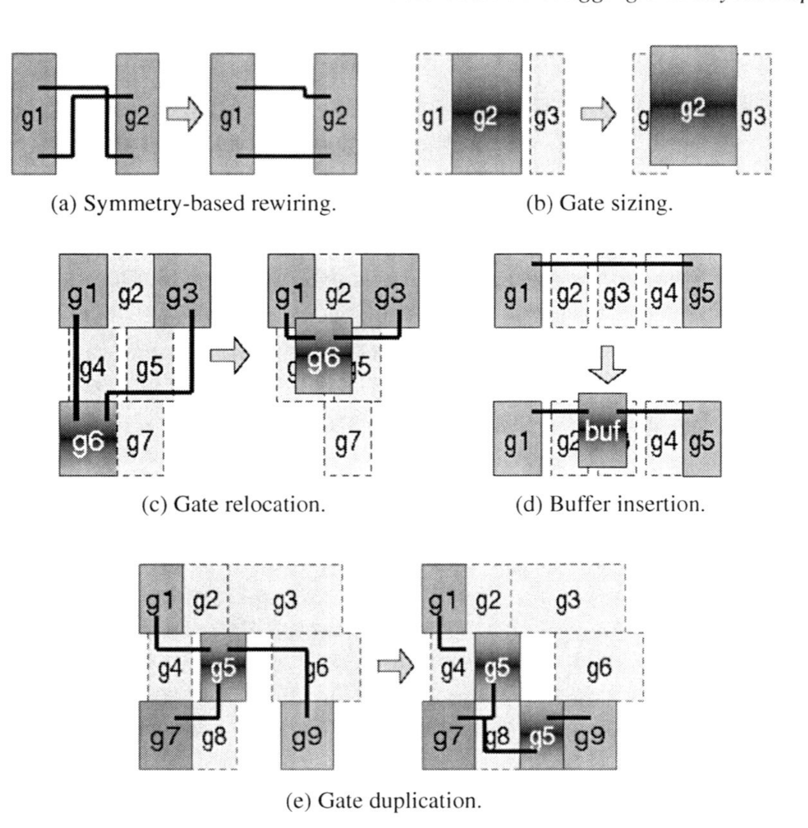

(a) Symmetry-based rewiring.

(b) Gate sizing.

(c) Gate relocation.

(d) Buffer insertion.

(e) Gate duplication.

Figure 11.1. Several distinct physical synthesis techniques. Newly-introduced overlaps are removed by legalizers after the optimization phase has completed.

powerful and how accurate the legalizer is, the physical parameters of the circuit may improve or deteriorate.

Traditional restructuring focuses on directing the synthesis process using timing information obtained from a placed or routed circuit. It is more aggressive in that it may change the logic structure as well as the placement. This technique reflects the fact that timing-driven synthesis requires accurate timing, which can only be obtained from a placed circuit. However, a circuit cannot be placed unless it is synthesized. Restructuring attempts to bridge the gap between these two different stages in circuit design.

A typical restructuring flow includes: (1) obtaining accurate timing analysis results from a placed or routed design, (2) identifying critical paths in the design, (3) selecting gates from the critical paths to form critical regions, (4) performing timing-driven resynthesis on the critical regions, and (5) calling legalizers to remove gate overlaps that may be created during the process. This process is repeated until timing closure is achieved. The work by Lu et al. [94],

Vaishnav et al. [129] and Changfan et al. [48] is all based on this flow with emphasis on different aspects. For example, the work by Vaishnav focuses on eliminating late-arriving events identified by symbolic simulation, while Changfan analyzes the effects of routing on timing and utilizes them in his resynthesis and incremental placement engines.

Traditional restructuring is usually physically unsafe. For example, evaluation of new cell locations cannot be done reliably for technology-independent restructuring unless technology mapping is also performed. Moreover, restructuring techniques based on AIGs are likely to be unsafe because node mergers performed in an AIG may distort a given placed circuit [148]. As a result, the effects of the changes are not immediately measurable. Although carefully designed techniques can be used to alleviate this problem [84, 91, 96], it is difficult to be eliminated altogether. The strength and safeness of these techniques are summarized in Table 11.1. The two physically safe techniques will be adapted to repair electrical errors in Section 11.4.

Table 11.1. Comparison of a range of physical synthesis techniques in terms of physical safeness and optimization potential.

Techniques	Physical safeness	Optimization potential
Symmetry-based rewiring	**Safe**	**Low**
SafeResynth	**Safe**	**Medium**
ATPG-based rewiring, buffer insertion, gate sizing, gate relocation	Unsafe*	Low
Gate replication	Unsafe*	Medium
Restructuring	Unsafe	High

Low potential means that only local optimizations are possible, and high potential means that large scale optimizations are possible. *Note: some of these techniques could be made safe but popular implementations use them in an unsafe fashion, allowing gate overlap.

11.2 New Resynthesis Technique – SafeResynth

Our safe physical synthesis approach, SafeResynth, is discussed in detail in this section. It uses *signatures* (see Section 5.1.1) produced by simulation to identify potential resynthesis opportunities, whose correctness are then validated by equivalence checking [148]. Since our goal is layout optimization and error repair, we can prune some of the opportunities based on their improvement potential before formally verifying them. To this end, we propose pruning techniques based on physical constraints and logical *compatibility* among signatures. SafeResynth is powerful in that it does not restrict resynthesis to small geometric regions or small groups of adjacent wires. It is safe because the produced placement is always legal and the effect can be evaluated immediately.

11.2.1 Terminology

A *signature* is a bit-vector of simulated values of a wire. Given the signature s_t of a wire w_t to be resynthesized, and a certain gate g_1, a wire w_1 with signature s_1 is said to be *compatible* with w_t if it is possible to generate s_t using g_1 with signature s_1 as one of its inputs. In other words, it is possible to generate w_t from w_1 using g_1. For example, if $s_1 = 1$, $s_t = 1$ and $g_1 = AND$, then w_1 is compatible with w_t using g_1 because it is possible to generate 1 on an AND's output if one of its inputs is 1. However, if $s_1 = 0$, then w_1 is not compatible with w_t using g_1 because it is impossible to obtain 1 on an AND's output if one of its inputs is 0 (see Figure 11.4).

A *controlling value* of a gate is the value that fully specifies the gate's output when applied to one input of the gate. For example, 0 is the controlling value for AND because when applied to the AND gate, its output is always 0 regardless of the value of other inputs. When two signatures are *incompatible*, that can often be traced to a controlling value in some bits of one of the signatures.

11.2.2 SafeResynth Framework

The SafeResynth framework is outlined in Figure 11.2, and an example is shown in Figure 11.3. In this section we illustrate how timing can be optimized; however, SafeResynth can also be used to optimize other circuit parameters or repair post-silicon errors. Initially, $library$ contains all the gates to be used for resynthesis. We first generate a signature for each wire by simulating certain input patterns. In order to optimize timing, $wire_t$ in line 2 will be selected from wires on the critical paths in the circuit. Line 3 restricts our search of potential resynthesis opportunities according to certain physical constraints, and lines 4–5 further prune our search space based on logical soundness. After a valid resynthesis option is found, we try placing the gate on various overlap-free sites close to a desired location in line 6 and check their timing improvements. In this process, more than one gate may be added if there are multiple sinks for $wire_t$, and the original driver of $wire_t$ may be replaced. In line 10 we remove redundant gates and wires that may appear because $wire_t's$ original driver may no longer drive any wire, which often initiates a chain of further simplifications.

11.2.3 Search-Space Pruning Techniques

In order to resynthesize a target wire ($wire_t$) using an n-input gate in a circuit containing m wires, the brute force technique needs to check $\binom{m}{n}$

1. Simulate patterns and generate a signature for each wire.
2. Determine $wire_t$ to be resynthesized and retrieve $wires_c$ from the circuit.
3. Prune $wires_c$ according to physical constraints.
4. Foreach $gate \in library$ with inputs selected from combinations of compatible wires $\in wires_c$.
5. Check if $wire_t$'s signature can be generated using $gate$ with its inputs' signatures. If not, try next combination.
6. If so, do restructuring using $gate$ by placing it on overlap-free sites close to the desired location.
7. If timing is improved, check equivalency. If not equivalent, try next combination of wires.
8. If equivalent, a valid restructuring is found.
9. Use the restructuring with maximum delay improvement for resynthesis.
10. Identify and remove gates and wires made redundant by resynthesis.

Figure 11.2. The SafeResynth framework.

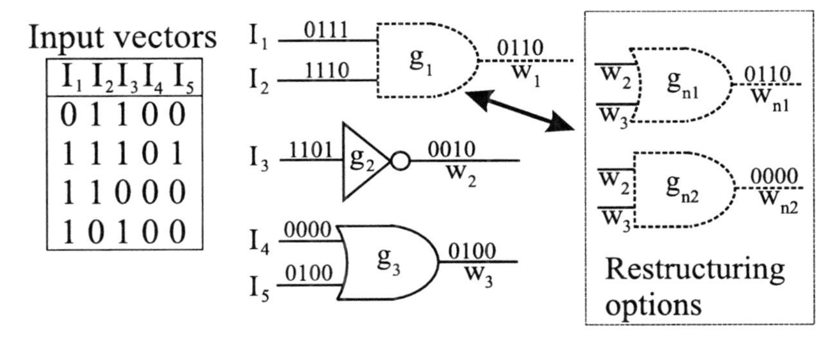

Figure 11.3. A restructuring example. Input vectors to the circuit are shown on the left. Each wire is annotated with its signature computed by simulation on those test vectors. We seek to compute signal w_1 by a different gate, e.g., in terms of signals w_2 and w_3. Two such restructuring options (with new gates) are shown as g_{n1} and g_{n2}. Since g_{n1} produces the required signature, equivalence checking is performed between w_{n1} and w_1 to verify the correctness of this restructuring. Another option, g_{n2}, is abandoned because it fails our compatibility test.

combinations of possible inputs, which can be very time-consuming for $n > 2$. Therefore it is important to prune the number of wires to try.

When the objective is to optimize timing, the following physical constraints can be used in line 3 of the framework: (1) wires with arrival time later than that of $wire_t$ are discarded because resynthesis using them will only increase delay; and (2) wires that are too far away from the sinks of $wire_t$ are abandoned because the wire delay will be too large to be beneficial. We set this distance threshold to twice the HPWL (Half-Perimeter WireLength) of $wire_t$.

In line 4 logical compatibility is used to prune the wires that need to be tried. Wires not compatible with $wire_t$ using $gate$ are excluded from our search.

Figure 11.4 summarizes how compatibilities are determined given a gate type, the signatures of $wire_t$ and the wire to be tested ($wire_1$).

Gate type	$wire_t$	$wire_1$	Result
NAND	0	0	Incompatible
NOR	1	1	Incompatible
AND	1	0	Incompatible
OR	0	1	Incompatible
XOR/XNOR	Any	Any	Compatible

Figure 11.4. Conditions to determine compatibility: $wire_t$ is the target wire, and $wire_1$ is the potential new input of the resynthesized gate.

To accelerate compatibility testing, we use the *one-count*, i.e., the number of 1s in the signature, to filter out unpromising candidates. For example, if $gate$==OR and the one-count of $wire_t$ is smaller than that of $wire_1$, then these two wires are incompatible because OR will only increase one-count in the target wire. This technique can be applied before bit-by-bit compatibility test to detect incompatibility faster.

Our *pruned_search* algorithm that implements lines 4–5 of the framework is outlined in Figure 11.5. It is specifically optimized for two-input gates and is a special version of the GDS algorithm shown in Section 6.3. $Wire_t$ is the target wire to be resynthesized, $wires_c$ are wires selected according to physical constraints, and $library$ contains gates used for resynthesis. All wires in the fanout cone of $wire_t$ are excluded in the algorithm to avoid the formation of combinational loops.

```
Function pruned_search(wire_t, wires_c, library)
1  foreach gate ∈ library
2     wires_g = compatible(wire_t, wires_c, gate);
3     foreach wire_1 ∈ wires_g
4        wires_d = get_potential_wires(wire_t, wire_1, wires_g, gate);
5        foreach wire_2 ∈ wires_d
6           restructure using gate, wire_1 and wire_2;
```

Figure 11.5. The pruned_search algorithm.

In Figure 11.5, function $compatible$ returns wires in $wires_g$ that are compatible with $wire_t$ given $gate$. Function $get\_potential\_wires$ returns wires in $wires_d$ that are capable of generating the signature of $wire_t$ using $gate$ and $wire_1$, and its algorithm is outlined in Figure 11.6. For XOR and XNOR, the signature of the other input can be calculated directly, and wires with signatures identical to that signature are returned using the signature hash table. For other gate types, signatures are calculated iteratively for each wire (denoted as

$wire_2$) using $wire_1$ as the other input, and the wires that produce signatures which match $wire_t's$ are returned.

```
Function get_potential_wires(wire_t, wire_1, wires_g, gate)
1   if (gate == XOR/XNOR)
2       wires_d= sig_hash[wire_t.signature XOR/XNOR wire_1.signature];
3   else
4       foreach wire_2 ∈ wires_g
5           if (wire_t.signature == gate.evaluate(wire_1.signature, wire_2.signature))
6               wires_d = wires_d ∪ wire_2;
7   return wires_d;
```

Figure 11.6. Algorithm for function get_potential_wires. XOR and XNOR are considered separately because the required signature can be calculated uniquely from $wire_t$ and $wire_1$.

11.3 Physically-Aware Functional Error Repair

In this section we describe our *Physically-Aware Functional Error Repair (PAFER)* framework, which is based on the CoRé framework (see Chapter 5). PAFER automatically diagnoses and fixes logic errors in the layout by changing its combinational portion. In this context, we assume that state values are available, and we treat connections to the flip-flops as primary inputs and outputs. To support the layout change required in logic error repair, we also develop a *Physically-Aware ReSynthesis (PARSyn)* algorithm.

11.3.1 The PAFER Framework

The algorithmic flow of our PAFER framework is outlined in Figure 11.7. The enhancements to make the CoRé framework physically-aware are marked in boldface. Note that unlike CoRé, the circuits (ckt_{err}, ckt_{new}) in PAFER now include layout information.

The inputs to the framework include the original circuit (ckt_{err}) and the test vectors ($vectors_p$, $vectors_e$). The output of the framework is a circuit (ckt_{new}) that passes verification and does not violate any physical constraints. In line 2 of the PAFER framework, the error is diagnosed, and the fixes are returned in $fixes$. Each fix contains one or more wires that are responsible for the circuit's erroneous behavior and should be resynthesized. In line 4 of the PAFER framework, $PARSyn$ is used to generate a set of new resynthesized circuits (ckt_{new}), which will be described in the next subsection. These circuits are then checked to determine if any physical constraint is violated. For example, whether it is possible to implement the change using metal fix. In lines 5–6, that no circuit complies with the physical constraints means no valid implementation can be found for the current fix. As a result, the fix will be abandoned and the next fix will be tried. Otherwise, the first circuit that does not violate any physical constraints is selected in line 7, where the circuits in

```
framework PAFER(ckt_err, vectors_p, vectors_e, ckt_new)
 1   calculate ckt_err's initial signatures using vectors_p and vectors_e;
 2   fixes = diagnose(ckt_err, vectors_e);
 3   foreach fix ∈ fixes
 4      ckts_new = PARSyn(fix, ckt_err);
 5      if (every circuit in ckts_new violates physical constraints)
 6         continue;
 7      ckt_new = the first circuit in ckts_new that does not violate physical constraints;
 8      counterexample = verify(ckt_new);
 9      if (counterexample is empty)
10         return (ckt_new);
11      else
12         if (check(ckt_err, counterexample) fails)
13            fixes = rediagnose(ckt_err, counterexample, fixes);
14      simulate counterexample and update ckt's signatures;
```

Figure 11.7. The algorithmic flow of the PAFER framework.

$ckts_{new}$ can be pre-sorted using important physical parameters such as timing, power consumption, or reliability. The functional correctness of this circuit is then verified as in the original CoRé framework. Please refer to Chapter 5 for more details on this part of the framework.

11.3.2 The PARSyn Algorithm

The resynthesis problem in post-silicon debugging is considerably different from traditional ones because the numbers and types of spare cells are often limited. As a result, traditional resynthesis flows may not work because technology mapping the resynthesis function using the limited number of cells can be difficult. Even if the resynthesis function can be mapped, implementing the mapped netlist may still be infeasible due to other physical limitations. Therefore, it is desirable in post-silicon debugging that the resynthesis technique can generate as many resynthesized netlists as possible.

To support this requirement, our PARSyn algorithm exhaustively tries all possible combinations of spare cells and input signals in order to produce various resynthesized netlists. To reduce its search space, we also develop several pruning techniques based on logical and physical constraints. Although exhaustive in nature, our PARSyn algorithm is still practical because the numbers of spare cells and possible inputs to the resynthesized netlists are often small in post-silicon debugging, resulting in a significantly smaller search space than traditional resynthesis problems.

Our PARSyn algorithm is illustrated in Figure 11.8, which tries to resynthesize every wire ($wire_t$) in the given fix. In line 2 of the algorithm, $getSpareCell$ searches for spare cells within $RANGE$ and returns the results in $spareCells$, where $RANGE$ is a distance parameter given by the engineer. This parameter limits the search of spare cells to those within $RANGE$ starting from $wire_t$'s driver. One way to determine $RANGE$ is to use the maximum length of a wire that FIB can produce. A subcircuit, ckt_{local}, is then extracted by $extractSubCkt$ in line 3. This subcircuit contains the cells which generate the signals that are allowed to be used as new inputs for the resynthesized netlists. A set of resynthesized netlists ($resynNets_{new}$) is then generated by $exhaustiveSearch$ in line 4. The cells in those netlists are then "placed" using spare cells in the layout to produce new circuits ($ckts_{new}$), which are returned in line 6.

```
function PARSyn(fix, ckt)
1  foreach wire_t ∈ fix
2     spareCells = getSpareCell(wire_t, ckt, RANGE);
3     ckt_local = extractSubCkt(wire_t, ckt, RANGE);
4     resynNets_new = exhaustiveSearch(1, spareCells, ckt_local);
5     ckts_new = placeResynNetlist(ckt, resynNets_new);
6  return (ckts_new);
```

Figure 11.8. The PARSyn algorithm.

To place the cells in a resynthesized netlist, we first sort spare cells according to their distances to $wire_t$'s driver. Next, we map each cell in the resynthesized netlist, the one closer to the netlist's output first, to the spare cell closest to $wire_t$'s driver. The reason behind this is that we assume the original driver is placed at a relatively good location. Since our resynthesized netlist will replace the original driver, we want to place the cell that generates the output signal of the resynthesized netlist as close to that location as possible. The rest of the cells in the resynthesized netlist are then placed using the spare cells around that cell.

The $exhaustiveSearch$ function called in the PARSyn algorithm is given in Figure 11.9. This function exhaustively tries combinations of different cell types and input signals in order to generate resynthesized netlists. The inputs to the function include the current logic level ($level$), available spare cells ($spareCells$), and a subcircuit (ckt_{local}) whose cells can be used to generate new inputs to the resynthesized netlists. The function returns valid resynthesized netlists in $netlists_{new}$.

In the function, $MAXLEVEL$ is the maximum depth of logic allowed to be used by the resynthesized netlists. So when $level$ equals to $MAXLEVEL$, no further search is allowed, and all the cells in ckt_{local} are returned (lines 1–2). In line 3, the search starts branching by trying every valid cell type, and the search is bounded if no spare cells are available for that cell type (lines 4–5). If a cell is available for resynthesis, it is deducted from the $spareCells$

```
function exhaustiveSearch(level, spareCells, ckt_local)
 1   if (level == MAXLEVEL)
 2      return all cells in ckt_local;
 3   foreach cellType ∈ validCellTypes
 4      if (checkSpareCell(spareCells, cellType) fails)
 5         continue;
 6      spareCells[cellType].count- -;
 7      netlists_sub = exhaustiveSearch(level + 1, spareCells, ckt_local);
 8      netlists_n = generateNewCkts(cellType, netlists_sub);
 9      netlists_n = checkNetlist(netlists_n, spareCells);
10      netlists_new = netlists_new ∪ netlists_n;
11   if (level == 1)
12      removeIncorrect(netlists_new);
13   return netlists_new;
```

Figure 11.9. The exhaustiveSearch function.

repository in line 6. In line 7 the algorithm recursively generates subnetlists for the next logic level, and the results are saved in $netlist_{sub}$. New netlists ($netlists_n$) for this logic level are then produced by $generateNewCkts$. This function produces new netlists using a cell with type=$cellType$ and inputs from combinations of subnetlists from the next logic level. In line 9 $checkNetlist$ checks all the netlists in $netlist_n$ and removes those that cannot be implemented using the available spare cells. All the netlists that can be implemented are then added to a set of netlists called $netlists_{new}$. If $level$ is 1, the logic correctness of the netlists in $netlists_{new}$ is checked by $removeIncorrect$, and the netlists that cannot generate the correct resynthesis functions will be removed. The rest of the netlists will then be returned in line 13. Note that BUFFER should always be one of the valid cell types in order to generate resynthesized netlists whose logic levels are smaller than MAXLEVEL. The BUFFERs in a resynthesized netlist can be implemented by connecting their fanouts to their input wires without using any spare cells.

To bound the search in $exhaustiveSearch$, we also used the logic pruning techniques described in Section 6.3. To further reduce the resynthesis runtime, we use netlist connectivity to remove unpromising cells from our search pool, e.g., cells that are too far away from the erroneous wire. In addition, cells in the fanout cone of the erroneous wire are also removed to avoid the formation of combinational loops.

11.4 Automating Electrical Error Repair

The electrical errors found post-silicon are usually unlikely to happen in any given region of a circuit, but become statistically significant in large chips. To this end, a slight modification of the affected wires has a high probability to successfully repair the problem. Being able to check this by performing accurate simulation and comparing several alternative fixes also increase the chances of successfully repairing the circuit even further. In this section we first describe two techniques that can automatically find a variety of electrical error repair options, including *SymWire* and *SafeResynth*. These techniques are able to generate layout transformations that modify the erroneous wires without affecting the circuit's functional correctness. Next, we study three cases to show how our techniques can be used to repair electrical errors.

11.4.1 The SymWire Rewiring Technique

Symmetry-based rewiring changes the connections between gates using symmetries. An example is illustrated in Figure 11.11(b), where the inputs to cells g_1 and g_2 are symmetric and thus can be reconnected without changing the circuit's functionality. The change in connections modifies the electrical characteristics of the affected wires and can be used to fix electrical errors. Since this rewiring technique does not perturb any cells, it is especially suitable for post-silicon debugging. In light of this, we propose an electrical error repair technique using the symmetry-based rewiring method presented in Chapter 7. This technique is called *SymWire* and its algorithm is outlined in Figure 11.10. The input to the algorithm is the wire (w) that has electrical errors, and this algorithm changes the connections to the wire using symmetries. In line 1, we extract various subcircuits ($subCircuits$) from the original circuit, where each subcircuit has at least one input connecting to w. Currently, we extract subcircuits composed of 1–7 cells in the fanout cone of w using breadth-first search and depth-first search. For each extracted subcircuit, which is saved in ckt, we detect as many symmetries as possible using function $symmetryDetect$ (line 3). If any of the symmetries involve a permutation of w with another input, we swap the connections to change the electrical characteristics of w. The symmetry detector can be implemented using the techniques presented earlier in Section 7.2.

```
function SymWire(w)
1   extract subCircuits with w as one of the inputs;
2   foreach ckt ∈ subCircuits
3     sym = symmetryDetect(ckt);
4     if (sym involves permutation of w with another input)
5       reconnect wires in ckt using sym;
```

Figure 11.10. The SymWire algorithm.

11.4.2 Adapting SafeResynth to Perform Metal Fix

Some electrical errors cannot be fixed by modifying a small number of wires, and a more aggressive technique is required. In this subsection we show

how the SafeResynth technique described in Section 11.2 can be adapted to perform post-silicon metal fix.

Assume that the error is caused by wire w or the cell g that drives w. We first use SafeResynth to find an alternative way to generate the same signal that drives w. In post-silicon debugging, however, we only rely on the *spare cells* that are embedded into the design but not connected to other cells. Therefore we do not need to insert new cells, which would be impossible to implement with metal fix. Next, we drive a portion or all of w's fanouts using the new cell. Since a different cell can also be used to drive w, we can change the electrical characteristics of both g and w in order to fix the error. Note that SafeResynth subsumes cell relocation; therefore, it can also find layout transformations involving replacements of cells.

11.4.3 Case Studies

In this subsection we show how our techniques can repair drive strength and coupling problems, as well as avoid the harm caused by the antenna effect. Note that these case studies only serve as examples, and our techniques can also be applied to repair many other errors.

Drive strength problems occur when a cell is too small to propagate its signal to all the fanouts within the designed timing budget. Our SafeResynth technique solves this problem by finding an alternative source to generate the same signal. As illustrated in Figure 11.11(a), the new source can be used to drive a fraction of the fanouts of the problematic cell, reducing its required driving capability.

Coupling between long parallel wires that are next to each other can result in delayed signal transitions under some conditions and also introduces unexpected signal noise. Our SafeResynth technique can prevent these undesirable phenomena by replacing the driver for one of the wires with an alternative signal source. Since the cell that generates the new signal will be at a different location, the wire topology can be changed. Alternatively, SymWire can also be used to solve the coupling problem. As shown in Figure 11.11(b), the affected wires no longer travel in parallel for long distances after rewiring, which can greatly reduce their coupling effects.

Antenna effects are caused by the charge accumulated during semiconductor manufacturing in partially-connected wire segments. This charge can damage and permanently disable transistors connected to such wire segments. In less severe situations, it changes the transistors' behavior gradually and reduces the reliability of the circuit. Because the charge accumulated in a metal layer will be eliminated when the next layer is processed, it is possible to split the total charge with another layer by breaking a long wire and going up or down one layer through vias. Based on this observation, *metal jumpers* [63] have been used to alleviate the antenna effect, where vias are intentionally inserted to

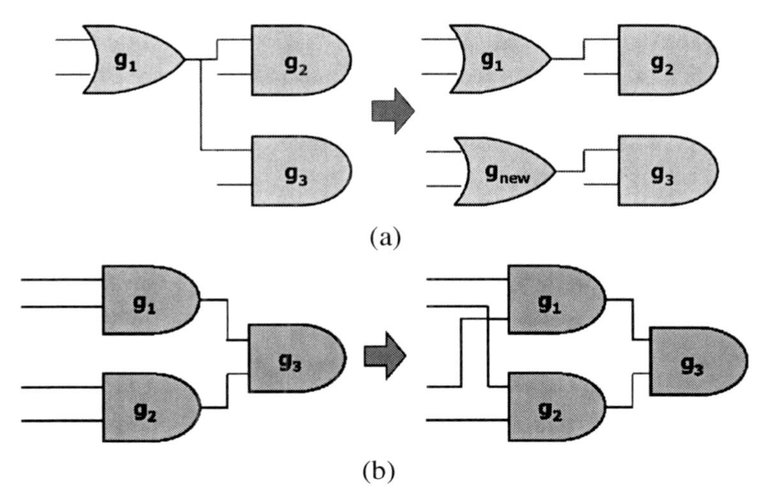

Figure 11.11. Case studies: (**a**) g_1 has insufficient driving strength, and SafeResynth uses a new cell, g_{new}, to drive a fraction of g_1's fanouts; (**b**) SymWire reduces coupling between parallel long wires by changing their connections using symmetries, which also changes metal layers and can alleviate the antenna effect (electric charge accumulated in partially-connected wire segments during the manufacturing process).

change layers for long wires. However, the new vias will increase the resistivity of the nets and slow down the signals. To this end, our SymWire technique can find transformations that change the metal layers of several wires to reduce their antenna effects. In addition, it allows simultaneous optimization of other parameters, such as the coupling between wires, as shown in Figure 11.11(b).

11.5 Experimental Results

To measure the effectiveness of the components in our FogClear post-silicon methodology, we conducted two experiments. In the first experiment we apply PAFER to repair functional errors in a layout; while the second experiment evaluates the effectiveness of SymWire and SafeResynth in finding potential electrical fixes. To facilitate metal fix, we pre-placed spare cells uniformly using the whitespace in the layouts, and they occupied about 70% of each layout's whitespace. These spare cells included INVERTERs, as well as two-input AND, OR, XOR, NAND, and NOR gates. In the PAFER framework, we set the $RANGE$ parameter to 50 μm and $MAXLEVEL$ to 2. Under these circumstances, approximately 45 spare cells (on average) are available when resynthesizing each signal. All the experiments were conducted on an AMD Opteron 880 workstation running Linux. The benchmarks were selected from OpenCores [154] except DLX, Alpha, and EXU_ECL. DLX and Alpha were from the Bug UnderGround project [149], while EXU_ECL was the control unit of OpenSparc's EXU block [164]. The characteristics of our benchmarks are summarized in Table 11.2. In the table, "#FFs" is the number of flip-flops and "#Cells" is the cell count of each benchmark. To produce the layouts for our experiments, we first synthesized the RTL designs with Cadence RTL Compiler 4.10 using a cell library based on the 0.18 μm technology node. We then placed the synthesized netlists with Capo 10.2 [24] and routed them with Cadence NanoRoute 4.10.

Table 11.2. Characteristics of benchmarks.

Benchmark	Description	#FFs	#Cells
Stepper	Stepper Motor Drive	25	226
SASC	Simple Asynchronous Serial Controller	117	549
EXU_ECL	OpenSparc EXU control unit	351	1460
Pre_norm	Part of FPU	71	1877
MiniRISC	MiniRISC full chip	887	6402
AC97_ctrl	WISHBONE AC 97 Controller	2199	11855
USB_funct	USB function core	1746	12808
MD5	MD5 full chip	910	13311
DLX	5-stage pipeline CPU running MIPS-Lite ISA	2062	14725
PCI_bridge32	PCI bridge	3359	16816
AES_core	AES Cipher	530	20795
WB_conmax	WISHBONE Conmax IP Core	770	29034
Alpha	5-stage pipeline CPU running Alpha ISA	2917	38299
Ethernet	Ethernet IP core	10544	46771
DES_perf	DES core	8808	98341

11.5.1 Functional Error Repair

To evaluate our PAFER framework, we chose several benchmarks and injected functional errors at either the gate level or the RTL. At the gate level we injected bugs that complied with Abadir's error model (see Section 5.1.4), while those injected at the RTL were more complex errors (DLX contained real bugs). We collected input patterns for the benchmarks from several traces generated by verification (some of the traces were reduced by Butramin), and a golden model was used to generate the correct output responses and state values for error diagnosis and correction. Note that the golden model can be a high-level behavior model because we do not need the simulation values for the internal signals of the circuit. The goal of this experiment was to fix the layout of each benchmark so that the circuit produces correct output responses

Table 11.3. Post-silicon functional error repair results.

Benchmark	Bug description	#Pat-terns	#Resyn. cells	Changes after repair			Runtime (sec)
				#Vias (%)	WL(%)	Delay(%)	
SASC(GL1)	Missing wire	90	2	0.29	1.27	–0.13	9.9
SASC(GL2)	Incorrect gate	66	1	0.13	0.33	0.00	4.4
EXU_ECL (GL1)	Incorrect gate	90	No valid fix was found				158.71
EXU_ECL (GL2)	Wrong wire	74	0	0.01	0.03	0.00	145.3
Pre_norm (GL1)	Incorrect wire	46	2	0.10	0.24	–0.05	38.92
DLX(GL1)	Incorrect gate	46	0	0.38	0.02	0.00	17245
DLX(GL2)	Additional wire	33	0	–0.13	–0.04	–0.15	12778
Pre_norm (RTL1)	Reduced OR replaced by reduced AND	672	3	0.19	0.38	0.57	76.24
MD5(RTL1)	Incorrect state transition	201	3	0.02	0.03	–0.02	29794
DLX(RTL1)	SLTIU inst. selects the wrong ALU operation	2208	No valid fix was found				12546
DLX(RTL2)	JAL inst. leads to wrong bypass from MEM stage	1536	0	0.00	0.00	0.03	8495
DLX(RTL3)	Incorrect forwarding for ALU+IMM inst.	1794	0	0.00	0.00	0.03	13807
DLX(RTL4)	Does not write to reg31	1600	No valid fix was found				7723
DLX(RTL5)	If RT = 7 memory write is incorrect	992	0	0.00	0.00	0.00	5771

The bugs in the upper half were injected at the gate level, while those in the lower half were injected at the RTL. Some errors can be repaired by simply reconnecting wires and do not require the use of any spare cell, as shown in Column 4.

for the given input patterns. This is similar to the situation described in Section 2.4 where fixing the observed errors allows the silicon die to be used for further verification. If the repaired die fails further verification, new counterexamples will be used to refine the fix as described in the PAFER framework. The results are summarized in Table 11.3, where "#Patterns" is the number of input patterns used in each benchmark, and "#Resyn. cells" is the number of cells used by the resynthesized netlist. In order to measure the effects of our fix on important circuit parameters, we also report the changes in via count ("#Vias"), wirelength ("WL"), and maximum delay ("Delay") after the layout is repaired. These numbers were collected after running NanoRoute in its ECO mode, and then they were compared to those obtained from the original layout. The maximum delay was reported by NanoRoute's timing analyzer.

The results in Table 11.3 show that our techniques can successfully repair logic errors for more than 70% of the benchmarks. We analyzed the benchmarks that could not be repaired and found that in those benchmarks, cells that produce the required signals were too far away and were excluded from our search. As a result, our resynthesis technique could not find valid fixes. In practice, this also means that the silicon die cannot be repaired via metal fix. The results also show that our error-repair techniques may change physical parameters such as via count, wirelength, and maximum delay. For example, the wirelength of SASC(GL1) increased by more than 1% after the layout was repaired. However, it is also possible that the fix we performed will actually improve these parameters. For example, the via count, wirelength, and maximum delay were all improved in DLX(GL2). In general, the changes in these physical parameters are typically small, showing that our error-repair techniques have few side effects.

11.5.2 Electrical Error Repair

We currently do not have access to tools that can identify electrical errors in a layout. Therefore, in this experiment we evaluate the effectiveness of our electrical error repair techniques by computing the percentages of wires where at least one valid transformation can be found. To this end, we selected 100 random wires from each benchmark and assumed that the wires contained electrical errors. Next, we applied SymWire and SafeResynth to find layout transformations that could modify the wires to repair the errors. The results are summarized in Table 11.4. In the table, "#Repaired" is the number of wires that could be modified, and "Runtime" is the total runtime of analyzing all 100 wires. We also report the minimum, maximum and average numbers of metal segments affected by our error-repair techniques. These numbers include the segments removed and inserted due to the layout changes.

From the results, we observe that both SymWire and SafeResynth were able to alter more than half of the wires for most benchmarks, suggesting that they can effectively find layout transformations that change the electrical characteristics of the erroneous wires. In addition, the number of affected metal segments is often small, which indicates that both techniques have little physical impact on the chip, and the layout modifications can be implemented easily by FIB. The runtime comparison between these techniques shows that SymWire runs significantly faster than SafeResynth because symmetry detection for small subcircuits is much faster than equivalence checking. However, SafeResynth is able to find and implement more aggressive layout changes for more difficult errors: as the results suggest, SafeResynth typically affects more metal segments than SymWire, producing more aggressive physical modifications. We also observe that SymWire seems to perform especially well for arithmetic cores such as MD5, AES_core, and DES_perf, possibly due to the

Table 11.4. Results of post-silicon electrical error repair.

Benchmark	SymWire					SafeResynth				
	#Re-paired	Metal seg. affected			Runtime (sec)	#Re-paired	Metal seg. affected			Runtime (sec)
		Min	Max	Mean			Min	Max	Mean	
Stepper	**81**	6	33	**15.7**	0.03	**79**	14	53	**28.3**	4.68
SASC	**50**	8	49	**19.8**	0.79	**41**	2	48	**27.8**	3.32
EXU_ECL	**68**	7	42	**15.0**	1.13	**71**	14	831	**119.1**	23.02
MiniRISC	**58**	4	29	**13.7**	1.65	**57**	14	50	**28.1**	166
AC97_ctrl	**52**	9	26	**13.9**	3.26	**56**	14	53	**31.9**	68.02
USB_funct	**70**	7	36	**16.4**	1.84	**58**	16	74	**32.4**	157.52
MD5	**82**	7	30	**15.0**	1.83	**79**	13	102	**37.9**	2630
DLX	**64**	6	49	**15.8**	11.00	**67**	13	97	**40.2**	8257
PCI_bridge32	**42**	8	42	**16.6**	6.04	**32**	15	54	**31.2**	211.28
AES_core	**83**	5	32	**15.0**	2.53	**83**	12	64	**31.4**	285.58
WB_conmax	**84**	7	35	**16.0**	2.96	**46**	19	71	**35.2**	317.50
Alpha	**67**	9	41	**16.3**	12.32	**55**	11	101	**36.9**	85104
Ethernet	**36**	7	22	**13.4**	45.01	**18**	18	104	**46.6**	3714
DES_perf	**91**	7	1020	**36.7**	4.86	**76**	10	60	**29.0**	585.34

100 wires were randomly selected to be erroneous, and "#Repaired" is the number of errors that could be repaired by each technique. The number of metal segments affected by each technique is also shown.

large numbers of logic operations used in these cores. Since many basic logic operations are symmetric (such as AND, OR, XOR), SymWire is able to find many repair opportunities. On the other hand, SymWire seems to perform poorly for benchmarks with high percentages of flip-flops, such as SASC, PCI_bridge32, and Ethernet. The reason is that SymWire is not able to find symmetries in flip-flops. As a result, if many wires only fanout to flip-flops, SymWire will not be able to find fixes for those wires.

11.6 Summary

Due to the dramatic increase in design complexity, more and more errors are escaping pre-silicon verification and are discovered post-silicon. While most steps in the IC design flow have been highly automated, little effort has been devoted to the post-silicon debugging process, making it difficult and ad hoc. To address this problem, we use our FogClear methodology to systematically automate the post-silicon debugging process, and it is powered by our new techniques and algorithms that enhance key steps in post-silicon debugging. The integration of logical, spatial and electrical considerations in these techniques facilitates the generation of netlists and layout transformations to fix the bug, and these techniques are complemented by search pruning methods for more scalable processing. These ideas form the foundation of our PAFER framework and the PARSyn algorithm that correct functional errors, as well as the SymWire and SafeResynth methods to repair electrical errors. Our empirical results show that these techniques can repair a substantial number of errors in most benchmarks, demonstrating their effectiveness for post-silicon debugging. FogClear can also reduce the costs of respins: fixes generated by FogClear only impact metal layers, hence enabling the reuse of transistor masks. The accelerated post-silicon debugging process also promises to shorten the time to the next respin, which can limit revenue loss due to late market entry.

Chapter 12

METHODOLOGIES FOR SPARE-CELL INSERTION

Post-silicon validation has recently become a major bottleneck in IC design. Several high profile IC designs have been taped-out with latent bugs, and forced the manufacturers to resort to additional design revisions. Such changes can be applied through metal fix; however, this is impractical without carefully pre-placed spare cells. As Figure 12.1(a) shows, a good spare-cell selection and placement facilitate metal fix with minimal perturbation of the silicon die. On the other hand, Figures 12.1(b) and (c) show that poorly placed spare cells can only be reached through long wires, leading to large increments in the propagation delay of the circuit; and that a poor selection of cell types requires the use of more cells to fix the same error.

In this chapter we perform comprehensive analysis of the issues related to spare-cell insertion, including the types of spare cells that should be used as well as their placement. In addition, we propose a new technique to measure the heterogeneity among signals and use it to determine spare-cell density. Finally, we integrate our findings into a multi-faceted approach that calculates

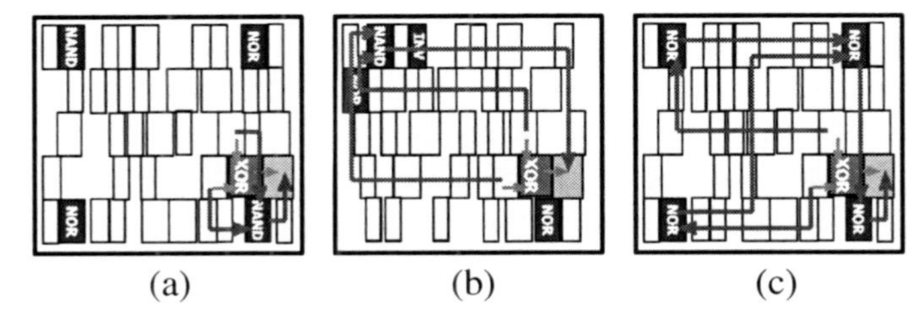

Figure 12.1. A design where an XOR gate must be replaced by a NAND using spare cells. (**a**) A high-quality fix with little perturbation of the layout. (**b**) A low-quality fix that requiring long wires due to poor spare-cell placement. (**c**) Another low-quality fix using several cells due to a poor selection of cell types.

regional demand for spare cells, identifies the most appropriate cell types, and places such cells into the layout.

Our work offers the first evaluation of strategies for spare-cell selection and placement in the context of post-silicon debugging. It answers the following important questions.

- What types of spare cells are most useful for metal fix?
- Do different types of designs or bugs need different combinations of spare cells? How to select such combinations?
- What is the impact of different spare-cell placement methods on important circuit parameters before and after metal fix?
- Should spare-cell density be different in different regions of a circuit? How to determine the best density automatically?

Key methods presented in this chapter are:

- A new technique to evaluate which type of cell is most useful to repair a given circuit, called *SimSynth*. SimSynth also measures the heterogeneity among signals and is the first technique that addresses the cell density problem.
- A spare-cell insertion methodology that covers both spare-cell selection and placement.

12.1 Existing Spare-Cell Insertion Methods

Spare-cell insertion is a design-dependent challenge whose solutions often rely on bug analysis from previous chips. Due to the confidentiality of relevant data, the results are only revealed in patents. Existing techniques for spare-cell insertion either provide better spare cells that are more powerful in generating new logic functions [17, 31, 66, 87, 105, 107, 119, 137, 131, 143] or strive to find better placement for the spare cells, so that they are located in proximity of a potential metal fix demand [17, 21, 31, 66, 87, 107, 131, 143]. Since these techniques are only described in patents, no empirical evaluation has been reported, particularly in the context of metal fix. As a result, their utility in post-silicon debug remains unclear. In Table 12.1 we summarize existing solutions that address the spare-cell insertion problem. Note that some techniques emphasize elevating lower-level wires for easier FIB access, which can also reduce respin cost because only masks for upper-level metal need to be updated. However, the elevated vias and metal segments may cause routing congestion and worsen circuit delay, hurting the overall circuit's performance in the end.

Table 12.1. A summary of existing spare-cell insertion techniques described in US patents. Major contributions are marked in boldface.

Author, year	Spare cell type	Placement and routing methods	Drawbacks/limitations
Yee [143], 1997	Most commonly used cell in the design; **one of the earliest works on spare-cell insertion**	Spare cells scattered after placement	Designed for 2 metal layers only
Lee [87], 1997	**NAND/NOR gates with many inputs, BUF, INV, DFF (new spare-cell selection)**	**Placed close to potentially buggy region**	High-input gates may waste space; other cell types may be more useful
Payne [107], 1999	**Gate array (new structure)**	Spare cells scattered after placement	No new placement technique claimed
Wong [137], 2001	**Configurable logic and INV (new structure)**	N/A	No placement technique claimed
Schadt [119], 2002	**Programmable cells (new structure); elevated lower-level wires improve FIB access**	Spare-cell islands scattered before placement	Uses 2 metal layers only; inputs/outputs of spare cells must be elevated
Chaise-martin [31], 2003	**NAND, DFF, trigate (new structure)**	Placed in a zigzag pattern; **stand-by tracks for routing**	Stand-by tracks may create routing congestion
Bingert [17], 2003	Gate-array islands	**Floorplaned with the design, then scattered uniformly**	Spare-cell islands may occupy too much space
Giles [66], 2003	**New spare-cell selection within cell islands including INV, DFF, MUX, AND, NAND, NOR and BUF**	**Placed according to design hierarchy**	Each module is allowed only one additional I/O; only fixed blocks supported
Or-Bach [105], 2004	**New FPGA-like structure**	N/A	Uses 3 metal layers only; no placement technique claimed
Vergnes [131], 2004	**New structure with functional input bus and an equation input bus**	**Placed with potentially buggy modules by hardwiring inputs of spare cells to signals in those modules**	Occupied routing tracks may create congestion
Brazell [21], 2006	N/A	**Whitespace allocated during Floorplanning; cells inserted after placement**	Spare cells occupy all remaining whitespace – impractical for modern layouts

12.2 Cell Type Analysis

As suggested by Table 12.1, many existing techniques seek better selections of spare-cell types so as to generate more complex logic functions for metal fix [31, 66, 87, 131, 143]. Here, one tries to avoid low-utility cells that waste valuable whitespace. A careful analysis of references suggests that none of the existing techniques vary the spare-cell selection throughout the design. However, intuitively it seems that different circuit blocks in the design may benefit best from different types of spare cells, and in general, some types may be more useful than others. In this work we developed a new algorithm, called SimSynth, that evaluates quantitatively the usefulness of a specific cell type, and we deployed on a range of designs to determine if the type of spare cells has a relevant impact on the quality of metal fix. Note that currently, we only consider combinational cells in SimSynth. The utility of sequential cells will depend on the sequential error repair technique being applied, which is a more sophisticated problem.

12.2.1 The SimSynth Technique

The SimSynth algorithm relies on a pool of input vectors for the circuit that can either be provided by a high-level simulator or acquired through a random selection. We then compute a signature for each internal circuit wire. These signatures can be thought of as partial truth tables that exclude all controllability don't-cares and they are the input to the SimSynth algorithm as indicated in the pseudocode of Figure 12.2. The algorithm's output is the success rate to generate a signature that already exists in the design region. To collect the signatures, we select a random wire, search for gates within 40 μm from the driver of the wire, and then retrieve the signatures from the outputs of those gates to form a signature pool. $SimSynth$ is then called using the signature pool as its input. Note that the 40 μm constraint is based on the observation that cells too far away will not be useful in metal fix because the wires that connect to them will be too long and will exhibit significant delay. In addition, FIB cannot generate long wires efficiently.

```
function SimSynth(candiSigs)
1   foreach cell ∈ spareCellTypes
2     foreach inputSigs ∈ combinations of signatures from candiSigs
3       sig ← cell.compute(inputSigs);
4       if (sig ∈ candiSigs)
5         success[cell]++;
6       count[cell]++;
7   return success/count;
```

Figure 12.2. The SimSynth algorithm.

Since we are measuring how easily an existing signal can be re-generated, the cell utility is useful for electrical error repair, which generates resynthesized netlists without modifying the circuit's logic functions. However, this technique can also measure the cell utility for functional error repair. The reason is that we are comparing signatures (partial truth-tables) of the signals. If two signals share the same signature, they must be functionally similar, but can differ on input vectors that have not been used to generate the signatures. This is similar to fixing functional errors: typically, a new signal that fixes a functional error is only slightly different from an existing one because most of the circuit's functions are already correct in post-silicon debugging. In general, more input vectors will bias the utility of spare-cell types toward fixing electrical errors because the generated signals will be closer to existing ones, while the selection will be biased toward functional errors when fewer vectors are used. To make SimSynth more relevant to studying functional errors, we can also consider signatures that are only slightly different from an existing one: generating a signature that is 1-bit different from an existing one can also be counted as a success. In practice, it is also possible that a fix requires a significant change to the circuit's functions. Implementing such a dramatic change, however, typically requires more complex resynthesized netlists involving large numbers of spare cells, which can make metal fix difficult or even impossible. In this work we do not discuss the utility of spare cells for fixing such extensive errors.

Further analysis shows that SimSynth can also be used to determine spare-cell density. The reason is that what SimSynth really measures is the heterogeneity among signals in the circuit. If the success rate is high, then the logic functions of the signals are similar, and generating a new signal that is close to any existing one should be easy. If the rate is low, then the functions of signals are quite different from each other, and generating a new signal using those signals would require more gates. This analysis is confirmed by our experimental results shown in Section 12.5.2.

EXAMPLE 12.1 *Figure 12.3 shows a SimSynth execution example using a full adder, where gate g1 should be XOR instead of OR. Two input vectors are used, producing a 2-bit signature for each wire. Suppose we want to measure the utility of cell types for the region indicated by the dashed line that contains two distinct signatures. SimSynth tries different cell types with different combinations of inputs (only 1 combination in this example) and measures the success rate to replicate an existing signature. The results on the right of the figure show that AND and XOR are more useful than NAND in this case. Note that the correct cell type to fix the bug can be successfully identified because signatures are only partial truth tables, which allow the identification of spare cells that can generate different signals. In general, additional input vectors will bias the cell-type selection towards the one that allows less function change.*

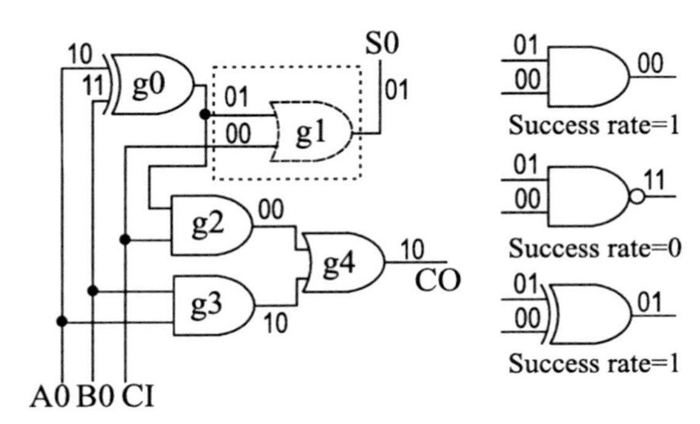

Figure 12.3. SimSynth example using a full adder.

12.2.2 Experimental Setup

Our implementation platform is based on the OAGear package [155] from Cadence Labs that uses the OpenAccess database and is integrated with the Capo placer [3]. We use benchmarks provided by the Bug UnderGround project [149] (Alpha), which includes a number of actual bugs found in fully functional microprocessor designs. Other benchmarks are from OpenCores [154] (MRISC, MD5 and DES_perf), picoJava (Hold_logic), and OpenSparc (EXU_ECL) [167]. The characteristics of these benchmarks are summarized in Table 12.2, where the first four are individual modules of the Alpha processor, followed by the full fledged Alpha design. Next, we show another processor (MRISC), followed by two CPU control blocks (Hold_logic and EXU_ECL) and two cryptographic cores (MD5 and DES_perf). To generate the layout information for these designs, we first synthesize the designs with Cadence RTL Compiler 4.10 based on a 0.18 μm library, and then we instruct Capo to place the design with uniform whitespace. By using uniform whitespace we produce lower bounds for the trends we observe, and the actual trends should be stronger with more realistic placement techniques that distribute design cells to aggressively optimize interconnect. We use Cadence NanoRoute 4.10 to route the final design and calculate the routed wirelength and circuit delay. The cell types considered in our analysis are INV (inverter), AND, OR, XOR, NAND, NOR, and MUX2. All cells except INV and MUX2 have two inputs. To evaluate a region with 200 signals using SimSynth, approximately 6 seconds are required on an AMD 2.4 GHz Opteron workstation.

Table 12.2. Characteristics of benchmarks

Benchmark	Description	Cell count	Delay (ns)
Alpha_IF	Instruction fetch unit of Alpha	1205	1.15
Alpha_ID	Instruction decode unit of Alpha	11806	1.91
Alpha_EX	Instruction execution unit of Alpha	20903	3.89
Alpha_MEM	Memory stage unit of Alpha	363	0.44
Alpha	Alpha CPU full chip	30212	6.93
MRISC	MiniRISC CPU	4359	2.66
Hold_logic	Part of PicoJava IU control	67	0.61
EXU_ECL	Part of OpenSparc EXU control	2083	0.99
MD5	MD5 encryption/decryption core	9181	6.92
DES_perf	DES encryption/decryption core	100776	3.37

12.2.3 Empirical Results

The experimental results are summarized in Figure 12.4, which shows two interesting trends. First, the distribution of cell-type utility varies widely among modules of the Alpha processor: signatures can often be re-generated easily using one gate in the IF and ID blocks, but not in the EX and MEM blocks. The reason is that IF and ID contain mostly control logic. Since control logic is mainly generated from "if-then" constructs, most signals are generated by ANDing, ORing or multiplexing the same group of signals. As a result, the logic functions between two signals are often very similar, making it easier to generate identical signatures using one gate. On the other hand, EX is dominated by datapaths. Since signals in such modules usually compute more distant functions, a single gate is less likely to re-generate an existing signature. For example, the first bit and the last bit in an adder compute very different functions. This result shows that to fix errors in arithmetic cores, more spare cells may be needed than fixing similar errors in control logic. Second, we observe that MUX2 is more useful in control logic (Alpha_IF, Alpha_ID, Hold_logic and EXU_ECL) than in arithmetic cores. The reason is that control logic is typically composed of many "if-then" constructs that can be efficiently implemented and modified using multiplexers.

12.2.4 Discussion

Our empirical results suggest that AND, NAND, OR, NAND and INV are the most useful in general, while XOR is the least useful. But CMOS standard cells that implement INV, NAND and NOR are smaller than those for AND and OR gates, making INV, NAND and NOR preferable as spare cells due to their functional completeness. The utility of MUX2, however, is unclear: it is

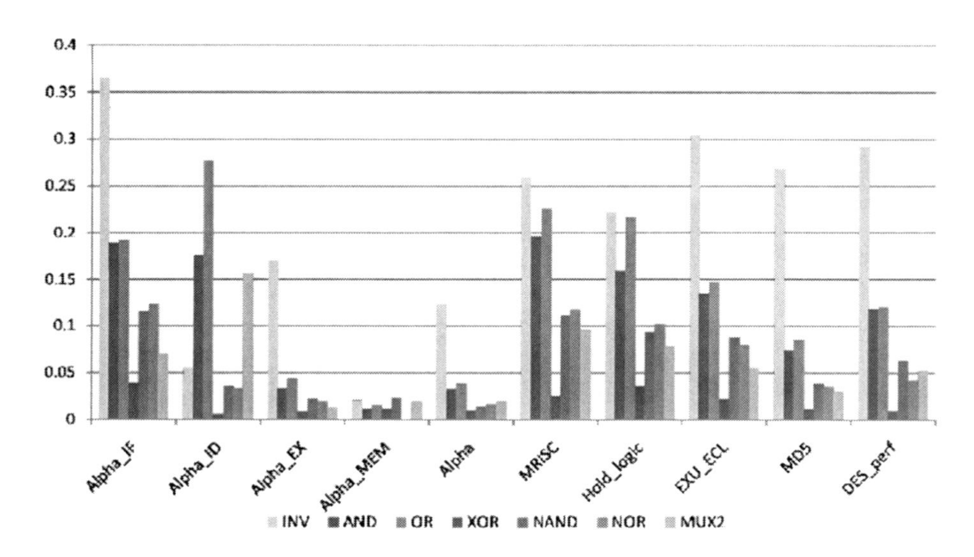

Figure 12.4. Using single gates of different types to generate desired signals. The success rates are shown in percent.

useful in only some of the benchmarks. Since MUX2 has three inputs, it should be useful in fixing functional errors because it can generate many different functions. In addition, the "if-then" construct commonly used in control logic can be modeled easily using MUX2. Since MUX2 is not a good candidate to fix electrical errors (MUX2 implemented using active transistors is large and slow), it should be implemented using pass transistors to fix functional errors.

In summary, our results suggest that: (1) different types of designs or errors need different combinations of spare-cell types; and (2) the most useful types are simple ones such as INV, NAND and NOR, while more complex gates such as XOR and MUX2 are less useful. Since there is no clear trend to predict the types of spare cells that will be more useful in a design, performing empirical analysis beforehand for each block in the design should help select the most adequate spare-cell types and distributions.

12.3 Placement Analysis

Placement of spare cells is another major factor that affects the quality of metal fix. When errors occur too far from pre-placed spare cells, the required wire connections may be too long to be practical. Even if such wires can be implemented by FIB or respin, the wire delay may also be large. Existing solutions either place the spare cells before design placement [31, 119], with design placement [66, 87, 131], or after design placement [17, 107, 143, 21]. To make sure that spare cells are available where necessary, uniform distribution of spare cells has been used by many existing solutions [17, 31, 119], while several other solutions focus on identifying potentially buggy regions and place spare cells close to them [66, 87, 131]. The spare cells are often grouped into spare-cell islands and then placed on a uniform grid; however, it is also possible to uniformly distribute individual cells instead of grouped cell islands. Since there is little research that evaluates different placement methods, the relative advantages of known techniques remain unclear.

A high-quality spare-cell placement should have minimal impact on important circuit parameters before metal fix to avoid increasing circuit delay or wirelength inadvertently and hurting design quality. It should also facilitate metal fix with the smallest impact on circuit parameters to provide high-quality repair. We observe that most existing techniques either scatter spare cells after design placement or place spare-cell islands uniformly before design placement. We call the former method PostSpare placement and the latter ClusterSpare. PostSpare covers the placement methods described in patents proposed by Yee [143] and Payne [107], while ClusterSpare covers those proposed by Schadt [119], Chaisemartin [31] and Bingert [17]. In ClusterSpare-based techniques, a cell island typically contains one cell for each selected type. Therefore, the number of cells in each island is usually large. An illustration of these placement methods is given in Figure 12.5.

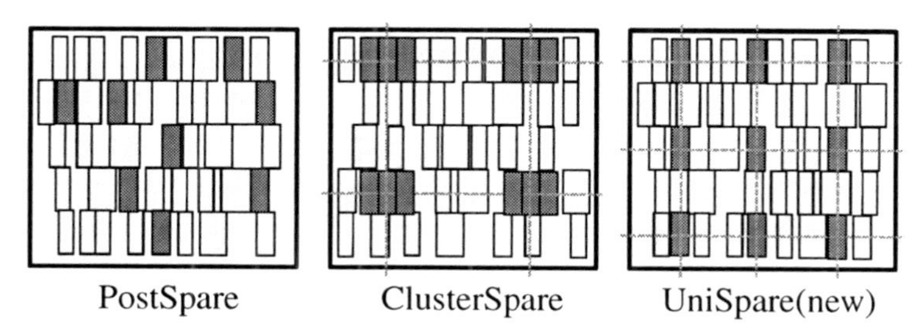

Figure 12.5. Illustration of different placement methods. Dark cells are spare cells. PostSpare inserts spare cells after design placement. Since design cells may be clustered in some regions, spare-cell distribution is typically non-uniform. ClusterSpare inserts spare-cell islands on a uniform grid before design placement, while UniSpare inserts single spare cells.

PostSpare placement should have minimal impact on important circuit parameters because spare cells are inserted after design placement. However, the error-repair quality of this method may be poor when design cells form high-utilization areas, forcing spare cells into sparser regions. When this happens, long wires may be needed to reach those cells. ClusterSpare placement may have larger impact on circuit parameters because the cell islands will act like macros and reduce the optimization that can be performed by the placer. However, it should provide better error-repair quality because their uniform distribution makes their access easier. Therefore, shorter wires can be used to reach the cell islands. In addition, connections among cells within the same island only require local wires and will be easy to implement. Note that, however, even the relatively short wires necessary to reach the spare-cell islands of ClusterSpare may trigger unacceptable wire delay increase in current silicon technology nodes, which are extremely delay-sensitive.

In this work we propose UniSpare, a solution that pre-places individual spare cells uniformly on a grid, as illustrated schematically in Figure 12.5(c). In this way, the average distance from a design cell to the closest spare cell is reduced. For example, when the size of the clusters reduces from 16 to 1 while maintaining the total number of spare cells, the average distance to reach a spare cell is reduced by 4 times. In a resynthesized netlist involving many gates, these individual cells can also act like buffers to increase signal strength, thus further reducing wire delay.

12.4 Our Methodology

Based on our analysis, we propose a new spare-cell insertion methodology, illustrated in Figure 12.6. Below we explain how it performs the selection and placement of spare cells.

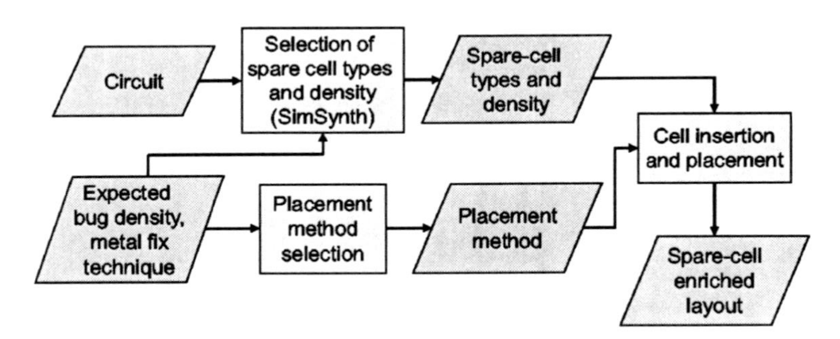

Figure 12.6. Our spare-cell insertion flow.

Our analysis suggests that different types of circuits require different distributions of spare-cell types. To select appropriate types, we apply our SimSynth technique in each design module and use the resulting cell-type distribution to determine the types of spare cells that should be inserted to each module. Since AND and OR gates require greater area than NAND and NOR, in our methodology we always use INV, NAND and NOR. In addition, for control blocks we insert multiplexers implemented using pass transistors to fix functional errors. Cell structures that provide greater flexibility, such as programmable logic or gate array [17, 105, 107, 119, 137], can also be used. However, they often require additional long wires to support programming.

The density of spare cells can be determined by the expected bug rate. If a circuit module is potentially buggy, then more spare cells should be placed in that module. For example, a perfectly working/verified circuit that is being scaled down to a new technology may encounter new electrical errors, but functional errors should not be prominent. In arithmetic cores, functional errors are relatively unlikely because these cores are usually heavily verified and are reused among designs. If bugs do occur, however, they may be difficult to repair using metal fix alone because all 32 or 64 bits may be affected. Wagner et al. [133] showed that most errors found in high-profile processors are in control logic. Therefore, more spare cells should be placed there.

If the expected bug rate is unknown, the results from SimSynth could be used. If the success rate measured by SimSynth in a block is lower than other blocks, then the heterogeneity among signals in the block is high and more spare cells should be placed. Suppose that there are n blocks in a circuit, the average success rate for block B_i is S_i, and the average success rate for all the blocks is S_{avg}. Also assume that the target overall spare-cell density is $D_{all}\%$. Formula 12.1 shows how to determine the spare-cell density D_i for block B_i. In the formula, P is a parameter that determines the impact of S_i on D_i and should be determined empirically. For example, based on our evaluation, P should be 20% for the blocks in the Alpha processor.

$$D_i = [\frac{(S_i - S_{avg}) \times P}{S_{avg}} + 1] \times \frac{D_{all}}{100\%} \tag{12.1}$$

The placement of spare cells depends on the expected bug rate and the metal-fix technique being used. If the expected bug rate is low, spare cells can be scattered uniformly after design placement. This helps ensure that spare cells do not affect circuit performance. If the expected bug rate is higher or unknown, then spare cells should be pre-placed uniformly before design placement so that wherever a fix must be applied, there are spare cells close to the repair site. To reduce the impact of the fix on important circuit parameters, spare cells should be placed individually or as small islands throughout the design using our proposed UniSpare method. Note that spare cells not connected during metal fix can also be used as buffers to improve circuit timing, as [51] suggests.

12.5 Experimental Results

In this section we empirically evaluate our techniques and compare them with existing solutions.

12.5.1 Cell-Type Selection

Experiment design: in this experiment we compare our results with two cell-selection methods: Giles [66] and Yee [143]. According to Figure 12.4, we use INV, NAND and NOR for most benchmarks, while Alpha_ID also includes

MUX2. Giles uses INV, DFF, MUX, AND, NAND, NOR and BUF as spare cells. Since Yee selects the "most-commonly used cell types" without indicating the number of types that should be used, we synthesized the benchmarks again using the seven types from which spare cells are drawn, and then selected the most-used two types for each benchmark, which were consistently NAND and INV. We use the UniSpare placement method for all three spare-cell selections to make sure the results are not affected by placement. To perform the experiment, we first select a subcircuit composed of 1–6 cells that are connected to each other. Next, we mimic a "fix" by resynthesizing the subcircuit and then map the resynthesized netlist to spare cells close to the subcircuit. Finally, we measure the delay and wirelength of the circuit after routing the modified netlist using NanoRoute's ECO mode. Better spare-cell selections should allow metal fix to be performed with smaller impact on circuit delay and wirelength. We ran each experiment 50 times to collect 50 data points for statistical analysis.

Results: the results are summarized in Figure 12.7. The graph shows that our spare-cell selection produces 23 and 4% smaller delay increase compared to Yee and Giles at a comparable wirelength increase. This result shows that our spare-cell selection can find more useful cells for each design and provides better error-repair quality after metal fix.

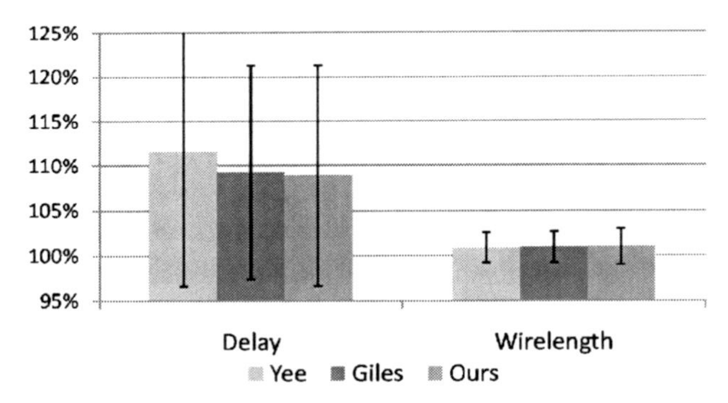

Figure 12.7. Delay and wirelength increase *after* metal fix when using three different sets of spare-cell selections. Ours has 23 and 4% smaller delay increase compared to Yee and Giles, while the wirelength increase is approximately the same.

12.5.2 Spare-Cell Placement

Three different types of placement methods are used in our experiments, and an illustration is given in Figure 12.5. PostSpare inserts individual spare cells after design placement; UniSpare inserts individual cells on a uniform grid before design placement; and ClusterSpare inserts spare-cell islands on a uniform grid before design placement, where each island is composed of 9 cells. We use INV, NAND and NOR gates as spare cells in our experiments, and each

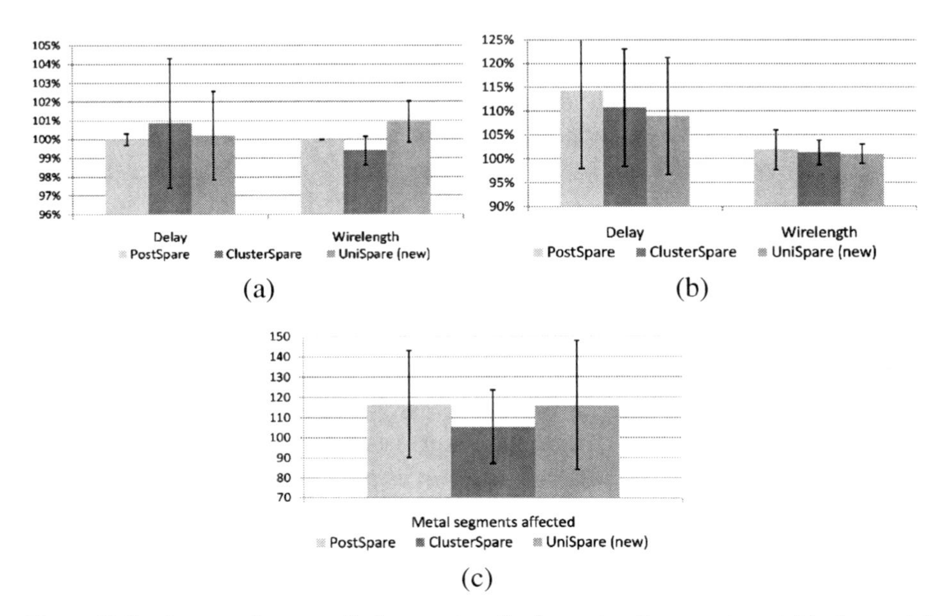

Figure 12.8. Impact of spare-cell placement methods on circuit parameters: (**a**) before metal fix; (**b**)(**c**) after metal fix. Ours has 24% smaller delay increase before metal fix compared with ClusterSpare. The delay increase after metal fix is 37 and 17% better than the PostSpare and ClusterSpare methods, respectively.

benchmark contains approximately 4% spare cells. The placer and router used in these experiments are Capo and NanoRoute. We ran each experiment 50 times to collect 50 distinct data points for statistical analysis.

Circuit parameter analysis before metal fix: in this experiment we first insert spare cells using the three methods described earlier. Next, we place and route the design using Capo and NanoRoute. Finally, we measure the impact of different placement methods on important circuit parameters, including delay and wirelength.

Figure 12.8(a) shows the average results of the benchmarks, and the error bars represent the range of one standard deviation. The figure shows that PostSpare placement does not affect circuit delay or wirelength. This is expected because the spare cells are placed after design placement; therefore, delay and wirelength should not be affected by spare-cell insertion. ClusterSpare placement shows a very interesting trend where the delay is increased while wirelength decreases. The reason is that large cell islands act like macros and force Capo to place design cells closer together, thus reducing total wirelength. At the same time, longer wires must be used to connect cells around the spare-cell islands, resulting in larger delay. For more aggressive placers, however, this trend may not be observed. The results also show that wirelength increased by 0.9% in UniSpare placement. This is because pre-placed spare cells will occupy certain placement sites, reducing the number of sites that can be used by the placer. Therefore, the optimization that can be performed by the placer will also be limited, resulting in larger wirelength. The delay, however, is only slightly affected by the inserted spare cells because connecting cells around a single cell only needs slightly longer wires, resulting in 24% smaller delay increase than ClusterSpare placement. We also note that the standard deviations are large in ClusterSpare and UniSpare placement methods, suggesting that spare-cell insertion may destabilize existing placement and routing tools.

Repair quality analysis after metal fix: after errors in a circuit have been repaired by metal fix, the circuit's major physical parameters may change, including interconnect length and maximum delay. Typically, repairs with higher quality can minimize the perturbation of those parameters. Since the quality of metal fix is affected by the placement of spare cells, we reused the experiment described in Section 12.5.1 to measure the impact of placement methods on error-repair quality. Since fixes that do not affect a critical path have no impact on circuit delay, we only selected data points whose delay has been changed to measure the true impact of placement methods on delay.

The average changes of physical parameters after metal fix are shown in Figure 12.8: Figure 12.8(b) shows the impact of placement on circuit delay and wirelength, while Figure 12.8(c) shows the impact on the number of affected metal segments. The error bars represent one standard deviation. The results show that PostSpare placement produces poor repair quality because it triggers a larger increase in delay and wirelength. In addition, it also affects more metal segments, making FIB more difficult. These trends should be stronger with non-uniform distribution of whitespace. From Figure 12.8(b), we observe that UniSpare placement has smaller delay and similar wirelength increase compared to ClusterSpare. The reason is that the cell islands placed by ClusterSpare are farther away from each other than the spare cells placed by UniSpare. As a result, longer wires are needed to connect to those cell islands, resulting in larger delay. On the other hand, Figure 12.8(c) shows that smaller numbers of metal segments are affected in circuits produced by ClusterSpare. This is because once those long wires reach the cell islands, connections among the cells in the same island only require local wires and will not perturb other wires. On average, UniSpare placement results in 37 and 17% smaller delay increase compared with PostSpare and ClusterSpare respectively, suggesting that it is the best placement method.

Density of spare cells: another interesting placement-related issue is the density of spare cells. Several existing techniques suggest that spare cells should be inserted close to potentially-buggy circuit modules [87, 131]. This approach is certainly useful if such information is available. However, it cannot be used if the bug distribution of a chip is unknown. As discussed in Section 12.2.1, SimSynth can address this problem. To evaluate its effectiveness, we counted the average number of spare cells used in the fixes produced by our previous experiment, and we contrast the results with Figure 12.4.

The results of this experiment are shown in Figure 12.9. This figure shows that to generate the same signal, the Alpha processor needs more spare cells than its EX block, followed by its ID and IF blocks. If we contrast this result with Figure 12.4, we can see that the IF block has the highest success rate in generating an existing signature using one spare cell, followed by ID, EX and the Alpha processor. These two observations are correlated because if it is easier to generate an existing signal using one gate, the number of cells needed to replicate a signal should also be smaller, at least on average. This phenomenon can also be observed on MD5 and DES_perf: MD5 requires more cells in each fix, and the success rate to generate an existing signal using one gate is also smaller. This result suggests that measuring the success rate of our SimSynth experiment can help determine the density of spare cells that should be placed on a silicon die.

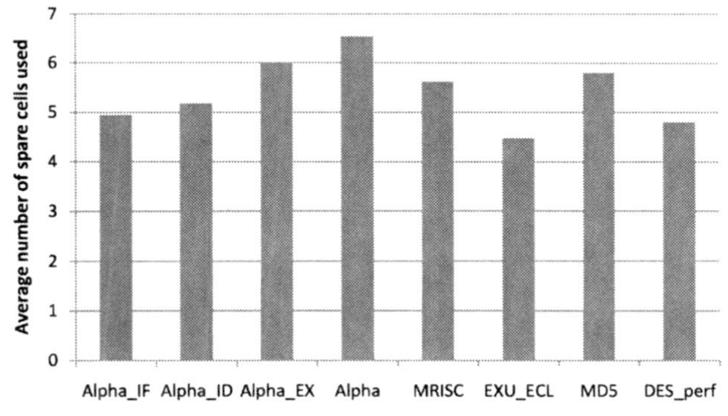

Figure 12.9. Average numbers of cells used when fixing bugs in the benchmarks. By contrasting with Figure 12.4 we show that SimSynth can help determine spare-cell density. For example, Alpha has smaller success rate in Figure 12.4 than its EX block, followed by its ID and IF blocks. This figure shows that the Alpha design requires more cells than its EX, ID and IF blocks.

12.6 Summary

In this chapter we performed a comprehensive analysis of spare-cell insertion to study the nature of this problem. Our work evaluates, for the first time, several rules of thumb commonly used in spare-cell insertion. First, several existing solutions suggest to use the "most-commonly used" cell type in the design as the spare-cell type. According to our results, the most popular cell type is indeed very useful, but (1) other types can be equally useful, and (2) using a blend of several spare-cell types provides better error-repair quality than using only one or two types. Second, most existing solutions use large spare-cell islands. Our analysis shows that this approach hurts circuit's wirelength and timing, and we believe that the difference will grow with each technology node due to poor scaling of interconnect delay. To reduce this impact, smaller islands should be used so as to reduce the average distance from a design cell to the closest spare cell. This will shorten the wires that connect to spare cells and improve circuit delay after metal fix. Third, most existing solutions neglect the impact of spare-cell insertion on circuit parameters. However, we showed that this impact may be significant. Without careful planning, spare-cell insertion can worsen circuit timing and wirelength.

Chapter 13

CONCLUSIONS

Verification is important in ensuring the correctness of a circuit design. As a result, it has been studied extensively and is highly automated. However, once errors are found, their diagnosis and correction are still mostly performed manually, which can be very difficult and time-consuming. Existing techniques that address this error-repair problem are often limited in their strength and scalability. This deficiency can be explained, in part, by the lack of scalable resynthesis methods. In addition, existing gate-level error-diagnosis techniques cannot be applied to the RTL, where most design activities occur, making automatic functional error correction much more difficult. This problem is further exacerbated by poor interoperability between verification and debugging tools, which stresses existing error-correction techniques even more. Since functional correctness is the most important aspect of high-quality designs, the resources consumed by debugging limit the effort that can be devoted to improve the performance of a circuit, hampering the sophistication of digital designs.

In this book we described several innovative algorithms, data structures, and methodologies that provide new ways for error diagnosis and correction. In addition, we described the FogClear framework that automates the functional error-repair process. This framework automatically corrects design errors at the RTL or gate level, and it is able to physically implement the corrections with minimal changes to existing cell locations, wire routes and manufacturing masks. In addition, our physical synthesis techniques are able to fix electrical errors with minimal impact to the layout. Below we summarize key techniques presented in this book.

- We presented a scalable bug trace minimizer, called *Butramin*, that reduces the complexity of bug traces and bridges the gap between verification and debugging.

- We described a *CoRé* resynthesis framework based on simulation and SAT. To achieve better scalability, we used an abstraction-refinement scheme in this framework. In addition, we devised a simplification of SPFDs, *Pairs of Bits to be Distinguished (PBDs)*, to encode the resynthesis information required by the framework. This representation supports the use of complete don't-cares and makes CoRé scale further than most existing error-correction techniques. Based on PBDs, we developed two innovative resynthesis algorithms, *Distinguishing-Power Search (DPS)* and *Goal-Directed Search (GDS)*, to support the logic changes required by error correction. The abstraction-refinement scheme in CoRé is conceptually different from those in existing solutions because CoRé's abstraction is based on signatures, which can be easily used by various resynthesis tools and extended to support different error-repair requirements. As articulated in Chapters 9 and 11, CoRé can be extended to repair RTL and post-silicon functional errors, but existing solutions do not have this flexibility. For example, it is difficult to utilize the abstraction proposed in [8] for automatic error correction.

- We designed a comprehensive and powerful functional symmetry detection algorithm for digital logic based on reduction to the graph-automorphism problem and available solvers. Given a multi-output logic function, this algorithm detects all symmetries of all known types, including permutations and phase-shifts on inputs and outputs, as well as the so-called higher-order symmetries. In addition, we devised a rewiring technique that uses the detected symmetries to optimize circuit wirelength or repair electrical errors discovered post-silicon.

- We introduced an innovative RTL error model that facilitates efficient and effective RTL error diagnosis. In addition, we proposed two diagnosis algorithms based on synthesis and symbolic simulation. Both techniques can scale much farther than existing gate-level diagnosis techniques, making our approach applicable to much larger designs. Our results also show that many more functional errors can be diagnosed compared with traditional gate-level diagnosis techniques.

- We outlined an incremental verification system, *InVerS*, that uses *similarity factor* to quickly estimate the functional correctness of physical synthesis optimizations. When errors are flagged, traditional verification techniques will be used. This system helps localize and identify bugs introduced by physical synthesis optimizations, and therefore decreases the risk from introducing new aggressive optimizations.

- We defined the concept of physical safeness and devised several physically safe techniques for post-silicon debugging. In order to repair functional errors, we proposed the *PAFER* framework and the *PARSyn* resynthesis algorithms. In addition, we illustrated two techniques, *SafeResynth* and *SymWire*, that can repair electrical errors on the layout. Since these techniques do not affect gate placements, they also allow metal fix. We also observed that spare-cell insertion plays an important role to the success of post-silicon metal fix. Therefore, we evaluated several spare-cell insertion methods and proposed new techniques to improve metal-fix quality.

- To facilitate comprehensive error repair at multiple stages of circuit design flow, we integrated several software components into a unified framework, called *FogClear*. This framework couples verification with debugging and can greatly reduce the debugging effort.

Empirical validation shows that all components of the FogClear framework are effective in performing their functions, and the integrated framework for post-silicon debugging is equally promising. With the help of FogClear, engineers will be able to diagnose and fix design errors more efficiently, which, we hope, will improve design quality and reduce cost.

References

[1] M. S. Abadir, J. Ferguson and T. E. Kirkland, "Logic Verification via Test Generation", *IEEE Trans. on Computer-Aided Design of Integrated Circuits and Systems*, Jan. 1988, pp. 138–148.

[2] M. Abramovici, P. Bradley, K. Dwarakanath, P. Levin, G. Memmi and D. Miller, "A Reconfigurable Design-for-Debug Infrastructure for SoCs", *Proc. Design Automation Conference (DAC)*, 2006, pp. 7–12.

[3] S. N. Adya, S. Chaturvedi, J. A. Roy, D. A. Papa and I. L. Markov, "Unification of Partitioning, Floorplanning and Placement", *Proc. International Conference on Computer-Aided Design (ICCAD)*, 2004, pp. 550–557.

[4] A. H. Ajami and M. Pedram, "Post-Layout Timing-Driven Cell Placement Using an Accurate Net Length Model with Movable Steiner Points", *Proc. Design Automation Conference (DAC)*, 2001, pp. 595–600.

[5] M. F. Ali, S. Safarpour, A. Veneris, M. S. Abadir and R. Drechsler, "Post-Verification Debugging of Hierarchical Designs", *Proc. International Conference on Computer-Aided Design (ICCAD)*, 2005, pp. 871–876.

[6] M. F. Ali, A. Veneris, S. Safarpour, R. Drechsler, A. Smith and M. S. Abadir, "Debugging Sequential Circuits Using Boolean Satisfiability", *Proc. International Conference on Computer-Aided Design (ICCAD)*, 2004, pp. 44–49.

[7] F. A. Aloul, A. Ramani, I. L. Markov and K. A. Sakallah, "Solving Difficult Instances of Boolean Satisfiability in the Presence of Symmetry", *IEEE Trans. on Computer-Aided Design of Integrated Circuits and Systems*, Sep. 2003, pp. 1117–1137.

[8] Z. S. Andraus, M. H. Liffiton and K. A. Sakallah, "Refinement Strategies for Verification Methods Based on Datapath Abstraction", *Proc. Asia and South Pacific Design Automation Conference (ASPDAC)*, 2006, pp. 19–24.

[9] Anonymous, "Conformal Finds DC/PhysOpt was Missing 40 DFFs!", ESNUG 464 Item 4, Mar. 30, 2007.

[10] K. Baker and J. V. Beers, "Shmoo Plotting: The Black Art of IC Testing", *IEEE Design and Test of Computers*, Vol. 14, No. 3, pp. 90–97, 1997.

[11] Z. Barzilai, J. L. Carter, and J. D. Rutledge, "HSS – a High-speed Simulator", *IEEE Trans. on Computer-Aided Design of Integrated Circuits and Systems*, Vol. 6, Jul. 1987, pp. 601–617.

[12] B. Bentley and R. Gray, "Validating the Intel Pentium 4 Processor", *Intel Technology Journal*, Q1 2001, pp. 1–8.

[13] Janick Bergeron, "Writing Testbenches: Functional Verification of HDL Models", Kluwer Academic Publishers, 2nd edition, 2003.

[14] V. Bertacco, "Scalable Hardware Verification with Symbolic Simulation", Springer, 2005.

[15] V. Bertacco and M. Damiani, "The Disjunctive Decomposition of Logic Functions", *Proc. International Conference on Computer-Aided Design (ICCAD)*, 1997, pp. 78–82.

[16] A. Biere, A. Cimatti, E. M. Clarke, and Y. Zhu, "Symbolic model checking without BDDs", in *Proc. International Conference on Tools and Algorithms for the Construction and Analysis of Systems (TACAS) – Lecture Notes in Computer Science (LNCS) 1579*, 1999, pp. 193–207.

[17] C. Bingert, C. D. Gorsuch, O. G. Mercado, A. K. Myers, J. A. Schadt and B. W. Yeager, "Integrated Circuit and Associated Design Method Using Spare Gate Islands", US Patent 6600341 B2, Jul. 2003.

[18] R. Bloem and F. Wotawa, "Verification and Fault Localization for VHDL Programs", *Journal of the Telematics Engineering Society (TIV)*, Vol. 2, 2002, pp. 30–33.

[19] D. Brand, R. A. Bergamaschi and L. Stok, "Be Careful with Don't Cares," *Proc. International Conference on Computer-Aided Design (ICCAD)*, 1995, pp. 83–86.

[20] R. E. Brayant, "Graph-Based Algorithms for Boolean Function Manipulation", *IEEE Trans. on Computers*, Aug. 1986, pp. 677–691.

[21] M. Brazell and A. Essbaum, "Method for Allocating Spare Cells in Auto-Place-Route Blocks", US Patent 6993738 B2, Jan. 2006.

[22] R. E. Brayant, D. Beatty, K. Brace, K. Cho and T. Sheffler, "COSMOS: a Compiled Simulator for MOS Circuits", *Proc. Design Automation Conference (DAC)*, 1987, pp. 9–16.

[23] M. L. Bushnell and V. D. Agrawal, *Essentials of Electronic Testing*, Kluwer, Boston, 2000.

[24] A. E. Caldwell, A. B. Kahng and I. L. Markov, "Can Recursive Bisection Alone Produce Routable Placements?", *Proc. Design Automation Conference (DAC)*, 2000, pp. 693–698.

[25] A. E. Caldwell, A. B. Kahng and I. L. Markov, "Toward CAD-IP Reuse: The MARCO GSRC Bookshelf of Fundamental CAD Algorithms", *IEEE Design and Test of Computers*, May 2002, pp. 72–81.

[26] D. V. Campenhout, H. Al-Asaad, J. P. Hayes, T. Mudge and R. B. Brown, "High-Level Design Verification of Microprocessors via Error Modeling", *ACM Trans. on Design Automation of Electronic Systems*, Oct. 1998, pp. 581–599.

[27] D. V. Campenhout, J. P. Hayes and T. Mudge, "Collection and Analysis of Microprocessor Design Errors", *IEEE Design and Test of Computers*, Oct.–Dec. 2000, pp. 51–60.

[28] D. Chai and A. Kuehlmann, "Building a Better Boolean Matcher and Symmetry Detector", *Proc. International Workshop on Logic and Synthesis (IWLS)*, Jun. 2005, pp. 391–398.

[29] D. Chai and A. Kuehlmann, "A Compositional Approach to Symmetry Detection in Circuits", *Proc. International Workshop on Logic and Synthesis (IWLS)*, Jun. 2006, pp. 228–234.

[30] D. Chai and A. Kuehlmann, "Symmetry Detection for Large Multi-Output Functions", *Proc. International Workshop on Logic and Synthesis (IWLS)*, May. 2007, pp. 305–311.

[31] P. Chaisemartin, "Structure and Method of Repair of Integrated Circuits", US Patent 6586961 B2, Jul. 2003.

[32] C. W. Chang, M. F. Hsiao, B. Hu, K. Wang, M. Marek-Sadowska, C. H. Cheng, and S. J. Chen, "Fast Postplacement Optimization Using Functional Symmetries", *IEEE Trans. on Computer-Aided Design of Integrated Circuits and Systems*, Jan. 2004, pp. 102–118.

[33] C. W. Chang and M. Marek-Sadowska, "Single-Pass Redundancy Addition and Removal", *International Conference on Computer-Aided Design (ICCAD)*, Nov. 2001, pp. 606–609.

[34] K.-H. Chang, V. Bertacco and I. L. Markov, "Simulation-Based Bug Trace Minimization with BMC-Based Refinement," *Proc. International Conference on Computer-Aided Design (ICCAD)*, Nov. 2005, pp. 1045–1051.

[35] K.-H. Chang, V. Bertacco and I. L. Markov, "Simulation-Based Bug Trace Minimization with BMC-Based Refinement," *IEEE Trans. on Computer-Aided Design of Integrated Circuits and Systems*, Jan. 2007, pp. 152–165.

[36] K.-H. Chang, I. L. Markov and V. Bertacco, "Post-Placement Rewiring and Rebuffering by Exhaustive Search For Functional Symmetries," *Proc. International Conference on Computer-Aided Design (ICCAD)*, Nov. 2005, pp. 56–63.

[37] K.-H. Chang, I. L. Markov and V. Bertacco, "Post-Placement Rewiring by Exhaustive Search For Functional Symmetries," *ACM Trans. on Design Automation of Electronic Systems*, Article 32, Aug. 2007, DOI=10.1145/1255456.1255469.

[38] K.-H. Chang, I. L. Markov and V. Bertacco, "Safe Delay Optimization for Physical Synthesis", *Proc. Asia and South Pacific Design Automation Conference (ASPDAC)*, Jan. 2007, pp. 628–633.

[39] K.-H. Chang, I. L. Markov and V. Bertacco, "SafeResynth: A New Technique for Physical Synthesis", *Integration: the VLSI Journal*, Jul. 2008, pp. 544–556.

[40] K.-H. Chang, I. L. Markov and V. Bertacco, "Fixing Design Errors with Counterexamples and Resynthesis", *Proc. Asia and South Pacific Design Automation Conference (ASPDAC)*, Jan. 2007, pp. 944–949.

[41] K.-H. Chang, I. L. Markov and V. Bertacco, "Fixing Design Errors with Counterexamples and Resynthesis", *IEEE Trans. on Computer-Aided Design of Integrated Circuits and Systems*, Jan. 2008, pp. 184–188.

[42] K.-H. Chang, I. L. Markov, and V. Bertacco, "Automating Post-Silicon Debugging and Repair", *Proc. International Conference on Computer-Aided Design (ICCAD)*, Nov. 2007, pp. 91–98.

[43] K.-H. Chang, I. L. Markov, and V. Bertacco, "Automating Post-Silicon Debugging and Repair", *IEEE Computer*, Vol. 41, No. 7, Jul. 2008, pp. 47–54.

[44] K.-H. Chang, D. A. Papa, I. L. Markov and V. Bertacco, "InVerS: An Incremental Verification System with Circuit Similarity Metrics and Error Visualization", *Proc. International Symposium on Quality Electronic Design (ISQED)*, Mar. 2007, pp.487–492.

[45] K.-H. Chang, I. Wagner, V. Bertacco and I. L. Markov, "Automatic Error Diagnosis and Correction for RTL Designs", *Proc. IEEE International High Level Design Validation and Test Workshop (HLDVT)*, Nov. 2007, pp. 65–72.

[46] K. H. Chang, I. L. Markov, and V. Bertacco, "Reap What You Sow: Spare Cells for Post-Silicon Metal Fix", *Proc. ACM International Symposium on Physical Design (ISPD)*, Mar. 2008, pp. 103–110.

[47] S. C. Chang, L. P. P. P. van Ginneken and M. Marek-Sadowska, "Circuit Optimization by Rewiring", *IEEE Trans. on Computers*, Sep. 1999, pp. 962–969.

[48] C. Changfan, Y. C. Hsu and F. S. Tsai, "Timing Optimization on Routed Designs with Incremental Placement and Routing Characterization", *IEEE Trans. on Computer-Aided Design of Integrated Circuits and Systems*, Feb. 2000, pp. 188–196.

[49] I. Chayut, "Next-Generation Multimedia Designs: Verification Needs", *Design Automation Conference (DAC)*, 2006, Section 23.2,
`http://www.dac.com/43rd/43talkindex.html`

[50] Y. A. Chen and F. S. Chen, "Algorithms for Compacting Error Traces", *Proc. Asia and South Pacific Design Automation Conference (ASPDAC)*, 2003, pp. 99–103.

[51] Y.-P. Chen, J.-W. Fang and Y.-W. Chang, "ECO Timing Optimization Using Spare Cells", *Proc. International Conference on Computer-Aided Design (ICCAD)*, pp. 530–535.

[52] C. Chiang and J. Kawa, "Design for Manufacturability and Yield for Nano-Scale CMOS", Springer, 2007.

[53] S.-J. Pan, K.-T. Cheng, J. Moondanos, Z. Hanna, "Generation of Shorter Sequences for High Resolution Error Diagnosis Using Sequential SAT", *Proc. Asia and South Pacific Design Automation Conference (ASPDAC)*, 2006, pp. 25–29.

[54] P.-Y. Chung and I. N. Hajj, "ACCORD: Automatic Catching and CORrection of Logic Design Errors in Combinational Circuits", *Proc. International Test Conference (ITC)*, 1992, pp. 742–751.

[55] J. Cong and W. Long, "Theory and Algorithm for SPFD-Based Global Rewiring", *Proc. International Workshop on Logic and Synthesis (IWLS)*, Jun. 2001, pp. 150–155.

[56] S. Cook, "Proc. ACM symposium on Theory of Computing", 1971, pp. 151–158.

[57] O. Coudert, C. Berthet and J. C. Madre, "Verification of Synchronous Sequential Machines Based on Symbolic Execution", *Proc. Automatic Verification Methods for Finite State Systems – Lecture Notes in Computer Science (LNCS) 407*, 1990, pp. 365–373.

[58] P. T. Darga, M. H. Liffiton, K. A. Sakallah, and I. L. Markov, "Exploiting Structure in Symmetry Detection for CNF", *Proc. Design Automation Conference (DAC)*, 2004, pp. 530–534.

[59] P. T. Darga, K. A. Sakallah, and I. L. Markov, "Faster Symmetry Discovery using Sparsity of Symmetries", *Proc. Design Automation Conference (DAC)*, 2008, pp. 149–154.

[60] R. A. Demillo, R. J. Lipton and F. G. Sayward, "Hints on Test Data Selection: Help for the Practicing Programmer", *IEEE Computer*, Apr. 1978, pp. 34–41.

[61] N. Eén and N. Sörensson, "An extensible SAT-solver", *Proc. Theory and Applications of Satisfiability Testing*, 2003, pp. 502–518.

[62] N. Eén and N. Sörensson, "Translating Pseudo-Boolean Constraints into SAT", *Journal on Satisfiability, Boolean Modeling and Computation (JSAT)*, 2006, pp. 1–25.

[63] J. Ferguson, "Turning Up the Yield", *IEE Electronics Systems and Software*, pp. 12–15, Jun./Jul. 2003.

[64] M. Gao, J. Jiang, Y. Jiang, Y. Li, S. Singha, and R. K. Brayton. MVSIS. *Proc. International Workshop on Logic and Synthesis (IWLS)*, 2001, `http://embedded.eecs.berkeley.edu/Respep/Research/mvsis/`

[65] P. Gastin, P. Moro, and M. Zeitoun, "Minimization of Counterexamples in SPIN", *Proc. SPIN – Lecture Notes in Computer Science (LNCS) 2989*, 2004, pp. 92–108.

[66] C. M. Giles, "Modular Collection of Spare Gates for Use in Hierarchical Integrated Circuit Design Process", US Patent 6650139 B1, Nov. 2003.

[67] R. Goering, "Post-Silicon Debugging Worth a Second Look", *EETimes*, Feb. 05, 2007.

[68] A. Groce and D. Kroening, "Making the Most of BMC Counterexamples", *Proc. Workshop on BMC*, 2004, pp. 71–84.

[69] R. Hildebrandt and A. Zeller, "Simplifying Failure-Inducing Input", *Proc. International Symposium on Software Testing and Analysis*, 2000, pp. 134–145.

[70] P.-H. Ho, T. Shiple, K. Harer, J. Kukula, R. Damiano, V. Bertacco, J. Taylor and J. Long, "Smart simulation using collaborative formal and simulation engines", *Proc. International Conference on Computer-Aided Design (ICCAD)*, 2000, pp. 120–126.

[71] M. Hrkic, J. Lillis and G. Beraudo, "An Approach to Placement-Coupled Logic Replication", *Proc. Design Automation Conference (DAC)*, 2004, pp. 711–716.

[72] A. Hu, "Formal Hardware Verification with BDDs: An Introduction", *Proc. Pacific Rim Conference (PACRIM)*, 1997, pp. 677–682.

[73] S.-Y. Huang, K.-C. Chen and K.-T. Cheng, "AutoFix: A Hybrid Tool for Automatic Logic Rectification", *IEEE Trans. on Computer-Aided Design of Integrated Circuits and Systems*, Sep. 1999, pp. 1376–1384.

[74] J.-H. R. Jiang and R. K. Brayton, "On the Verification of Sequential Equivalence", *IEEE Transactions on Computer-Aided Design*, Jun. 2003, pp. 686–697.

[75] T.-Y. Jiang, C.-N. J. Liu and J.-Y. Jou, "Estimating Likelihood of Correctness for Error Candidates to Assist Debugging Faulty HDL Designs", *Proc. International Symposium on Circuits and Systems (ISCAS)*, 2005, pp. 5682–5685.

[76] W. Jiang, T. Marwah and D. Bouldin, "Enhancing Reliability and Flexibility of a System-on-Chip Using Reconfigurable Logic", *Proc. International Symposium on Circuits and Systems (ISCAS)*, 2005, pp. 879–882.

[77] H. Jin, K. Ravi, and F. Somenzi, "Fate and Free Will in Error Traces", *Proc. International Conference on Tools and Algorithms for the Construction and Analysis of Systems (TACAS) – Lecture Notes in Computer Science (LNCS) 2280*, 2002, pp. 445–459.

[78] D. Josephson, "The Manic Depression of Microprocessor Debug", *Proc. International Test Conference (ITC)*, 2002, pp. 657–663.

[79] D. Josephson, "The Good, the Bad, and the Ugly of Silicon Debug", *Proc. Design Automation Conference (DAC)*, 2006, pp. 3–6.

[80] L. N. Kannan, P. R. Suaris and H. G. Fang, "A Methodology and Algorithms for Post-Placement Delay Optimization", *Proc. Design Automation Conference (DAC)*, 1994, pp. 327–332.

[81] K. Killpack, C. V. Kashyap and E. Chiprout, "Silicon Speedpath Measurement and Feedback into EDA Flows", *Proc. Design Automation Conference (DAC)*, 2007, pp. 390–395.

[82] A. Kolbl, J. Kukula and R. Damiano, "Symbolic RTL simulation", *Proc. Design Automation Conference (DAC)*, 2001, pp. 47–52.

[83] A. Kolbl, J. Kukula, K. Antreich and R. Damiano, "Handling Special Constructs in Symbolic Simulation", *Proc. Design Automation Conference (DAC)*, 2002, pp. 105–110.

[84] V. N. Kravets and P. Kudva, "Implicit Enumeration of Structural Changes in Circuit Optimization", *Proc. Design Automation Conference (DAC)*, 2004, pp. 438–441.

[85] A. Kuehlmann and F. Krohm, "Equivalence Checking Using Cuts and Heaps", *Proc. Design Automation Conference (DAC)*, 1997, pp. 263–268.

[86] S.-Y. Kuo, "Locating Logic Design Errors via Test Generation and Don't-Care Propagation", *Proc. European Design Automation Conference (EDAC)*, 1992, pp. 466–471.

[87] D. Lee, "Method and Apparatus for Quick and Reliable Design Modification on Silicon", US Patent 5696943, Dec. 1997.

[88] C. E. Leiserson and J. B. Saxe, "Retiming Synchronous Circuitry", *Algorithmica*, 1991, Vol. 6, pp. 5–35.

[89] N. G. Leveson and C. S. Turner, "An Investigation of the Therac-25 Accidents", *IEEE Computer*, Jul. 1993, pp. 18–41.

[90] D. M. Lewis, "A Hierarchical Compiled-code Event-driven Logic Simulator", *IEEE Trans. on Computer-Aided Design of Integrated Circuits and Systems*, Jun. 1991, pp. 726–737.

[91] C. Li, C-K. Koh and P. H. Madden, "Floorplan Management: Incremental Placement for Gate Sizing and Buffer Insertion", *Proc. Asia and South Pacific Design Automation Conference (ASPDAC)*, 2005, pp. 349–354.

[92] C.-C. Lin, K.-C. Chen and M. Marek-Sadowska, "Logic Synthesis for Engineering Change", *IEEE Trans. on Computer-Aided Design of Integrated Circuits and Systems*, Mar. 1999, pp.282–202.

[93] C.-H. Lin, Y.-C. Huang, S.-C. Chang and W.-B. Jone, "Design and Design Automation of Rectification Logic for Engineering Change", *Proc. Asia and South Pacific Design Automation Conference (ASPDAC)*, 2005, pp. 1006–1009.

[94] A. Lu, H. Eisenmann, G. Stenz and F. M. Johannes, "Combining Technology Mapping with Post-Placement Resynthesis for Performance Optimization", *Proc. International Conference on Computer Design (ICCD)*, 1998, pp. 616–621.

[95] F. Lu, M. K. Iyer, G. Parthasarathy, L.-C. Wang, and K.-T. Cheng and K.C. Chen, "An efficient sequential SAT solver with improved search strategies", *Proc. Design Automation and Test in Europe (DATE)*, 2005, pp. 1102–1107.

[96] T. Luo, H. Ren, C. J. Alpert and D. Pan, "Computational Geometry Based Placement Migration", *Proc. International Conference on Computer-Aided Design (ICCAD)*, 2005, pp. 41–47.

[97] J. C. Madre, O. Coudert and J. Pl. Billon, "Automating the Diagnosis and the Rectification of Design Errors with PRIAM", *International Conference on Computer-Aided Design (ICCAD)*, 1989, pp. 30–33.

[98] J. P. Marques-Silva and K. A. Sakallah, "GRASP: A Search Algorithm for Propositional Satisfiability", *IEEE Trans. on Computers*, Vol. 48, No. 5, May. 1999, pp. 506–521.

[99] J. Melngailis, L. W. Swanson and W. Thompson, "Focused Ion Beams in Semiconductor Manufacturing", *Wiley Encyclopedia of Electrical and Electronics Engineering*, Dec. 1999.

[100] A. Mishchenko, "Fast Computation of Symmetries in Boolean Functions", *IEEE Trans. on Computer-Aided Design of Integrated Circuits and Systems*, Nov. 2003, pp. 1588–1593.

[101] A. Mishchenko, J. S. Zhang, S. Sinha, J. R. Burch, R. Brayton and M. Chrzanowska-Jeske, "Using Simulation and Satisfiability to Compute Flexibilities in Boolean Networks", *IEEE Trans. on Computer-Aided Design of Integrated Circuits and Systems*, May 2006, pp. 743–755.

[102] D. Moller, J. Mohnke and M. Weber, "Detection of Symmetry of Boolean Functions Represented by ROBDDs", *Proc. International Conference on Computer-Aided Design (ICCAD)*, 1993, pp. 680–684.

[103] M. W. Moskewicz, C. F. Madigan, Y. Zhao, L. Zhang and S. Malik, "Chaff: Engineering an Efficient SAT Solver", *Proc. Design Automation Conference (DAC)*, 2001, pp. 530–535.

[104] D. Nayak and D. M. H. Walker, "Simulation-Based Design Error Diagnosis and Correction in Combinational Digital Circuits", *Proc. VLSI Test Symposium (VTS)*, 1999, pp. 25–29.

[105] Z. Or-Bach, "Customizable and Programmable Cell Array", US Patent 6756811 B2, Jun. 2004.

[106] S. Panda, F. Somenzi and B. F. Plessier, "Symmetry Detection and Dynamic Variable Ordering of Decision Diagrams", *Proc. International Conference on Computer-Aided Design (ICCAD)*, 1994, pp. 628–631.

[107] R. L. Payne, "Cell-Based Integrated Circuit Design Repair Using Gate Array Repair Cells", US Patent 5959905, Sep. 1999.

[108] S. M. Plaza, I. L. Markov and V. Bertacco, "Random Stimulus Generation using Entropy and XOR constraints", *Proc. Design Automation and Test in Europe (DATE)*, 2008, pp. 664–669.

[109] I. Pomeranz and S. M. Reddy, "On Determining Symmetries in Inputs of Logic Circuits", *IEEE Trans. on Computer-Aided Design of Integrated Circuits and Systems*, Nov. 1994, pp. 1428–1434.

[110] P. Rashinkar, P. Paterson and L. Singh, "System-on-a-chip Verification: Methodology and Techniques", Kluwer Academic Publishers, 2002.

[111] J.-C. Rau, Y.-Y. Chang and C.-H. Lin, "An Efficient Mechanism for Debugging RTL Description", *Proc. International Workshop on System-on-Chip for Real-Time Applications (IWSOC)*, 2003, pp. 370–373.

[112] K. Ravi and F. Somenzi, "High-Density Reachability Analysis", *Proc. International Conference on Computer-Aided Design (ICCAD)*, 1995, pp. 154–158.

[113] K. Ravi and F. Somenzi, "Minimal Satisfying Assignments for Bounded Model Checking", *Proc. International Conference on Tools and Algorithms for the Construction and Analysis of Systems (TACAS) – Lecture Notes in Computer Science (LNCS) 2988*, 2004, pp. 31–45.

[114] R. Rudell and A. Sangiovanni-Vincentelli, "Multiple-Valued Minimization for PLA Optimization", *IEEE Trans. on Computer-Aided Design of Integrated Circuits and Systems*, pp. 727–750, Sep. 1987.

[115] S. Safarpour and A. Veneris, "Trace Compaction using SAT-Based Reachability Analysis", *Proc. Asia and South Pacific Design Automation Conference (ASPDAC)*, 2007, pp. 932–937.

[116] S. Safarpour and A. Veneris, "Abstraction and Refinement Techniques in Automated Design Debugging", *Proc. Design Automation and Test in Europe (DATE)*, 2007, pp. 1182–1187.

[117] S. Safarpour, H. Mangassarian, A. Veneris, M. H. Liffiton and K. A. Sakallah, "Improved Design Debugging Using Maximum Satisfiability", *Proc. (FMCAD)*, 2007, pp. 13–19.

[118] S. Saranqi, S. Narayanasamy, B. Carneal, A. Tiwarj, B. Calder and J. Torrellas, "Patching Processor Design Errors with Programmable Hardware", *IEEE Micro*, Vol. 27, No. 1, 2007, pp. 12–25.

[119] J. A. Schadt, "Integrated Circuit with Standard Cell Logic and Spare Gates", US Patent 6404226 B1, Jun. 2002.

[120] N. Shenoy and R. Rudell, "Efficient Implementation of Retiming", *Proc. International Conference on Computer-Aided Design (ICCAD)*, 1994, pp. 226–233.

[121] S. Shen, Y. Qin, and S. Li, "A Fast Counterexample Minimization Approach with Refutation Analysis and Incremental SAT," *Proc. Asia and South Pacific Design Automation Conference (ASPDAC)*, 2005, pp. 451–454.

[122] C.-H. Shi and J.-Y. Jou, "An Efficient Approach for Error Diagnosis in HDL Design", *Proc. International Symposium on Circuits and Systems (ISCAS)*, 2003, pp. 732–735.

[123] S. Sinha, "SPFDs: A New Approach to Flexibility in Logic Synthesis", *Ph.D. Thesis*, University of California, Berkeley, May 2002.

[124] K. Shimizu and D. L. Dill, "Deriving a Simulation Input Generator and a Coverage Metric From a Formal Specification", *Proc. Design Automation Conference (DAC)*, 2002. pp. 801–806.

[125] A. Smith, A. Veneris and A. Viglas, "Design Diagnosis Using Boolean Satisfiability", *Proc. Asia and South Pacific Design Automation Conference (ASPDAC)*, 2004, pp. 218–223.

[126] S. Staber, B. Jobstmann and R. Bloem, "Finding and Fixing Faults", *Proc. Advanced Research Working Conference on Correct Hardware Design and Verification Methods (CHARME) – Lecture Notes in Computer Science (LNCS) 3725*, 2005, pp. 35–49.

[127] S. Staber, G. Fey, R. Bloem and R. Drechsler, "Automatic Fault Localization for Property Checking", *Lecture Notes in Computer Science (LNCS) 4383*, 2007, pp. 50–64.

[128] G. Swamy, S. Rajamani, C. Lennard and R. K. Brayton, "Minimal Logic Re-Synthesis for Engineering Change", *Proc. International Symposium on Circuits and Systems (ISCAS)*, 1997, pp. 1596–1599.

[129] H. Vaishnav, C. K. Lee and M. Pedram, "Post-Layout Circuit Speed-up by Event Elimination", *Proc. International Conference on Computer Design (ICCD)*, 1997, pp. 211–216.

[130] A. Veneris and I. N. Hajj, "Design Error Diagnosis and Correction via Test Vector Simulation", *IEEE Trans. on Computer-Aided Design of Integrated Circuits and Systems*, Dec. 1999, pp. 1803–1816.

[131] A. Vergnes, "Spare Cell Architecture for Fixing Design Errors in Manufactured Integrated Circuits", US Patent 6791355 B2, Sep. 2004.

[132] I. Wagner, V. Bertacco and T. Austin, "StressTest: An Automatic Approach to Test Generation via Activity Monitors", *Proc. Design Automation Conference (DAC)*, 2005. pp. 783–788.

[133] I. Wagner, V. Bertacco and T. Austin, "Shielding Against Design Flaws with Field Repairable Control Logic", *Proc. Design Automation Conference (DAC)*, 2006. pp. 344–347.

[134] D. E. Wallace, "Recognizing Input Equivalence in Digital Logic", *Proc. International Workshop on Logic and Synthesis (IWLS)*, 2001, pp. 207–212.

[135] G. Wang, A. Kuehlmann and A. Sangiovanni-Vincentelli, "Structural Detection of Symmetries in Boolean Functions", *Proc. International Conference on Computer Design (ICCD)*, 2003, pp. 498–503.

[136] C. Wilson and D. L. Dill, "Reliable Verification Using Symbolic Simulation with Scalar Values", *Proc. Design Automation Conference (DAC)*, 2000, pp. 124–129.

[137] J. Wong, D. Chiang and J. Tolentino, "Efficient Use of Spare Gates for Post-Silicon Debug and Enhancements", US Patent 6255845 B1, Jul. 2001.

[138] Y. L. Wu, W. Long and H. Fan, "A Fast Graph-Based Alternative Wiring Scheme for Boolean Networks", *Proc. International VLSI Design Conference*, 2000, pp. 268–273.

[139] Q. Wu, C. Y. R. Chen and J. M. Aken, "Efficient Boolean Matching Algorithm for Cell Libraries", *Proc. International Conference on Computer Design (ICCD)*, 1994, pp. 36–39.

[140] H. Xiang, L.-D. Huang, K.-Y. Chao and M. D. F. Wong, "An ECO Algorithm for Resolving OPC and Coupling Capacitance Violations", *Proc. International Conference On ASIC (ASICON)*, 2005, pp. 784–787.

[141] S. Yamashita, H. Sawada and A. Nagoya, "SPFD: A New Method to Express Functional Flexibility", *IEEE Trans. on Computer-Aided Design of Integrated Circuits and Systems*, pp. 840–849, Aug. 2000.

[142] Y.-S. Yang, S. Sinha, A. Veneris and R. E. Brayton, "Automating Logic Rectification by Approximate SPFDs", *Proc. Asia and South Pacific Design Automation Conference (ASPDAC)*, 2007, pp. 402–407.

[143] C. L. Yee, S. Aji and S. Rusu, "Method and Apparatus to Distribute Spare Cells within a Standard Cell Region of an Integrated Circuit", US Patent 5623420, Apr. 1997.

[144] J. Yuan, K. Albin, A. Aziz and C. Pixley, "Constraint Synthesis for Environment Modeling in Functional Verification", *Proc. Design Automation Conference (DAC)*, 2003, pp. 296–299.

[145] Q. K. Zhu and P. Kilze, "Metal Fix and Power Network Repair for SOC", *IEEE Computer Society Annual Symposium on VLSI (ISVLSI)*, 2006, pp. 33–37.

[146] `http://en.wikipedia.org/wiki/Pentium_FDIV_bug`

[147] `http://www.avery-design.com/`

[148] Berkeley Logic Synthesis and Verification Group, ABC: A System for Sequential Synthesis and Verification, Release 51205.
http://www-cad.eecs.berkeley.edu/~alanmi/abc/

[149] Bug UnderGround, http://bug.eecs.umich.edu/

[150] http://www.cadence.com/

[151] http://www.denali.com/

[152] "AMD Exec Discusses Barcelona Debacle", EE Times, Dec. 10, 2007

[153] International Technology Roadmap for Semiconductors 2005 Edition, http://www.itrs.net

[154] http://www.opencores.org/

[155] http://openedatools.si2.org/oagear/

[156] http://vlsicad.eecs.umich.edu/BK/PlaceUtils/

[157] ITC'99 Benchmarks(2nd release), http://www.cad.polito.it/tools/itc99.html

[158] UMICH Physical Design Tools, http://vlsicad.eecs.umich.edu/BK/PDtools/

[159] http://www.dafca.com/

[160] http://www.intel.com/design/core2duo/documentation.htm

[161] http://iwls.org/iwls2005/benchmarks.html

[162] http://www.openedatools.org/projects/oagear/

[163] http://www.synopsys.com/

[164] http://opensparc-t1.sunsource.net/

[165] picoJave Core, http://www.sun.com/microelectronics/picoJava/

[166] Pueblo SAT Solver, http://www.eecs.umich.edu/ hsheini/pueblo/

[167] http://www.sun.com/

[168] Verilog Simulator Benchmarks, http://www.veripool.com/verilog_sim_benchmarks.html

Index

Printed in the United States
143144LV00003BC/4/P

9 781402 093647